CHROMATOGRAPHIC SYSTEMS

maintenance and troubleshooting

second edition

ACADEMIC PRESS RAPID MANUSCRIPT REPRODUCTION

CHROMATOGRAPHIC SYSTEMS

maintenance and troubleshooting

second edition

JOHN Q. WALKER
mcdonnell-douglas research laboratory
st. louis, missouri

MINOR T. JACKSON, JR.
waters associates
st. louis, missouri

JAMES B. MAYNARD
mcdonnell-douglas materials laboratory
st. louis, missouri

ACADEMIC PRESS INC.
NEW YORK SAN FRANCISCO LONDON 1977
A Subsidiary of Harcourt Brace Jovanovich, Publishers

ACADEMIC PRESS, INC.
111 Fifth Avenue, New York, New York 10003

United Kingdom Edition published by
ACADEMIC PRESS, INC. (LONDON) LTD.
24/28 Oval Road, London NW1

Library of Congress Cataloging in Publication Data

Walker, John Q.
Chromatographic systems.

Includes index.
1. Chromatographic analysis—Apparatus and supplies—Maintenance and repair. I. Jackson, Minor T., joint author. II. Maynard, James B., joint author. III. Title.
QD117.C5W35 1977 681′.754 77-8356
ISBN 0-12-732052-0

PRINTED IN THE UNITED STATES OF AMERICA

80 81 82 9 8 7 6 5 4 3

CONTENTS

PART II — GAS CHROMATOGRAPHY

PREFACE TO THE SECOND EDITION

After writing the first edition, many new and useful ideas have been generated through discussions with several hundred chromatographers involved in American Chemical Society short courses on maintenance and troubleshooting. Most of these ideas are of primary importance to chromatographic instrumentation and have been described in this second revised edition. It also reflects the updating of liquid and gas chromatographic systems since 1971.

1977

John Q. Walter
Minor T. Jackson, Jr.
James B. Maynard

PREFACE TO THE FIRST EDITION

The purpose of this book is to provide a clear and concise guide for chromatographic maintenance—troubleshooting and repair procedures which can be utilized by both experienced and inexperienced chemists and technicians to reduce instrument down-time. No attempt is made to duplicate any of the excellent texts already published which may deal, in part, with troubleshooting chromatographic systems. Rather, we endeavor to go further by bridging the void between the chromatographer and the service engineer.

The text is divided into two parts. Liquid Chromatography, presented in Part I, consists of an introductory chapter on principles, techniques, and utility. This is followed by specific chapters devoted to the individual systems comprising the total liquid chromatographic makeup. The final liquid chromatography chapter, Comprehensive Troubleshooting, is provided for rapid reference. Part II, Gas Chromatography, begins with an introductory chapter on basic theory. This is followed by a systematic progression through possible malfunctions in various parts of the gas chromatograph, beginning with the sample introduction system.

The collaboration of three experienced chromatographers has definite advantages when considering diverse and complex subjects such as the maintenance and troubleshooting of chromatographic systems. The authors represent a total of 42 years of experience in several areas of chromatography and in different industrial backgrounds, i.e., aerospace, chemical manufacturing, and petroleum refining. For the past 10 years, each author has been involved in areas of chromatographic research with regard to the nonroutine applications associated with the needs of his employers. Included in these areas are those of control laboratory and analytical research supervision, plant start-up monitoring, and the usual day-to-day maintenance and troubleshooting problems. During our years of experience in various chromatographic functions, the authors have compiled careful notes regarding instrument malfunctions encountered in virtually every possible situation. Essentially, the text consists of these notes organized and rewritten in a readable, straightforward fashion.

The typical chromatographer who knows basic gas and liquid chromatographic theory and uses his equipment for many applications frequently considers the diagnosing and subsequent repair of instrument malfunctions something of an undesirable task and, in many cases, a black art. This need not be the situation because any adept chromatographer possesses the capability of demanding and receiving the maximum performance from his equipment. This capability is rewarding to him professionally and renders to him virtually full control and responsibility for the results attained. As a consequence, judgment regarding the validity of data is enhanced. In no context should this philosophy be construed that the chromatographer's primary goal be that of a service engineer. However, due to the advancements in chemical instrumentation technology, it behooves us to respect and appreciate some degree of inbreeding and/or overlap to fulfill our responsibilities as chromatographic scientists.

The operating cost of a $5000 chromatograph fully utilized for eight hours is approximately $45; however, the depreciation plus the overhead cost for the unit when not being fully utilized may well double the cost. This cost may be attributed to a reduction in the flow of analytical data, and, in many cases, less than full utilization of the analyst's time. A rapid diagnosis of the instrument or column malfunction with the appropriate remedy would substantially decrease instrument down-time and economize chromatographic maintenance.

Recent associations with various ASTM and API committees have shown that the gas chromatography expert must now be more proficient in the various types and capabilities of liquid chromatographic techniques in order to be a balanced separations scientist. So it is to this ''balanced separations scientist'' that we dedicate this book on the maintenance and the troubleshooting of chromatographic systems.

1972

John Q. Walker
Minor T. Jackson, Jr.
James B. Maynard

ACKNOWLEDGMENTS

The completion of this book would have been impossible without the help and constructive criticism of many people, especially Dr. R. P. W. Scott, Dr. M. P. T. Bradley, Mr. W. K. Hinrichs, and Mr. J. L. McDonald.

Grateful acknowledgment is due our managers—Dr. D. P. Ames and Dr. C. J. Wolf of McDonnell-Douglas Corporation, Mr. D. R. Beasecker and Dr. W. E. Koerner of Monsanto Company, and Dr. J. W. Armstrong of Shell Oil Company—for their helpful suggestions and ready assistance.

We wish to thank Mrs. Eunice Crayne, Mrs. Loretta Goins, and Mrs. Joan Waid for their efforts in preparing the manuscript, and Mr. R. H. Pfeffer and Mrs. Katherine Sewell for preparing the artwork.

Finally, special thanks to our wives, Virginia Walker, Pat Jackson, and Susan Maynard, and our children for their help, encouragement, and understanding during the past year.

CHROMATOGRAPHIC SYSTEMS

maintenance and troubleshooting

second edition

Part I

LIQUID CHROMATOGRAPHY

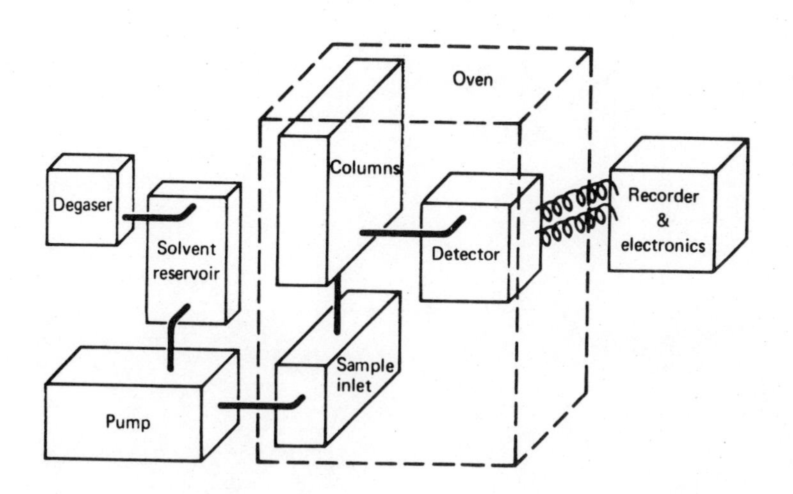

The general format of the liquid chromatography (LC) section of the text will be directed toward devoting specific chapters, in progression, to the individual systems comprising the LC make-up. The order in which these will be presented is as follows: Solvent Transport System, Sample Introduction System, Columns and Column Ovens, and Detection Systems. A final chapter on Comprehensive Troubleshooting is included also for rapid reference.

Chapter 1

CHROMATOGRAPHIC SYSTEMS: MAINTENANCE AND TROUBLESHOOTING

INTRODUCTION TO LIQUID CHROMATOGRAPHY

This portion of the text, dedicated to high-speed liquid chromatography, is presented as a guide to enable the chemist to recognize, maintain and/or eliminate malfunctions which may occur within his particular instrument via preventive maintenance and trouble-shooting techniques. Figure 1-1 shows a block diagram of a typical liquid chromatographic system.

Liquid chromatography, the oldest form of chromatography, has been in existence for many years and few analytical techniques, if any, are as powerful or offer the potential magnitude of its application. Until recently the technique was time-consuming, requiring as long as twenty-four hours to perform a single analysis. Currently, the same analysis is being performed in minutes and therefore, is deserving of its name, "high-speed liquid chromatography." This decrease in analysis time, with no subsequent loss in resolution, has been accomplished via efficient high-pressure pumping systems, increased column technology, high sensitivity detection systems and the overall reduction of chromatographic principles and knowledge to practice. Although much progress has been made, the present systems are far from optimum, and thorough knowledge of the technique and full awareness of its future potential cannot be completely visualized today. Virtually all parts of current

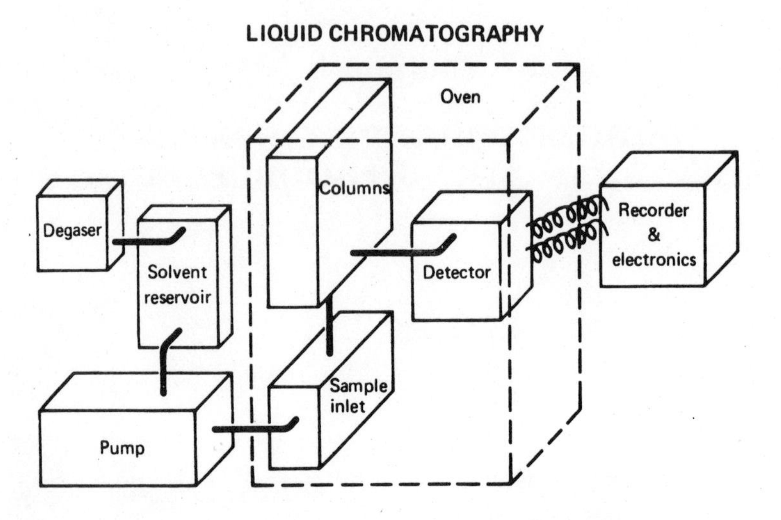

Fig. 1-1. Block diagram of a typical liquid chromatograph.

liquid chromatographic systems are being improved. It must, nonetheless, be realized that many problems associated with the present technology arise, at least in part, from insufficient knowledge and/or inadequate operation by the user.

There are excellent texts (1), (2), (7), (8), currently available devoted to theoretical and experimental considerations; however, some general background in liquid chromatography will be presented here relating to general principles and nomenclature. The four basic types of liquid chromatography may be categorized as follows:

1. Liquid-Solid or Adsorption Chromatography is an affinity separation associated with the separation of relatively nonpolar, hydrophobic materials. Normal phase adsorption consists of a non-polar solvent and a polar support. Silica gel and alumina are common choices using the solvents pentane and tetrahydrofuran, respectively. Silica gel is used to separate acidic or neutral compounds. Many workers note the undesirable effect of alumina in catalytically decomposing many organics; however, neutral aluminas are available which will completely eliminate or greatly minimize this problem. The solvent system is an equally important variable in adsorption chromatography. For a particular material, the more polar the solvent, the shorter the analysis time. Table 1-1 (3) consists of a eluotropic solvent

TABLE 1-1

Eluotropic solvent series (3)

Guide for solvent selection listed in order of increasing eluting power

Solvent	Dielectric Constant	Temperature
n-Hexane	1.9	(20°C)
Cyclohexane	2.0	(20°C)
Carbon Tetrachloride	2.2	(20°C)
Benzene	2.3	(20°C)
Toluene	2.4	(25°C)
Trichloroethylene	3.4	(16°C)
Diethyl ether	4.3	(20°C)
Chloroform	4.8	(20°C)
Ethyl acetate	6.0	(25°C)
1-Butanol	17.0	(25°C)
1-Propanol	20.0	(25°C)
Acetone	21.0	(25°C)
Ethanol	24.0	(25°C)
Methanol	33.0	(25°C)
Water	80.0	(20°C)

INCREASING ELUTING POWER

series and provides a convenient guide for solvent selection.

Thin-layer chromatography is also a type of adsorption chromatography and has proven to be an invaluable technique for screening various solvents prior to instrumental adsorption chromatography. In many cases, the binding material used in TLC plates prevents direct comparison of TLC and instrumental adsorption chromatography. Speed and efficiency of analysis, in addition to easier quantitative analysis and preparative applications, render the instrumental technique more practical than conventional thin-layer chromatography.

2. Liquid-Liquid or Partition Chromatography involves two liquid phases (one mobile, the other stationary). Preferably, the two liquids are immisible and exhibit different polarity. This technique is quite useful for homolog separations and can be used for virtually any separations involving components of the same or similar polarities. One inherent difficulty with this form of chromatography is that the mixture to be analyzed must exhibit solubility in the immiscible pair (mobile and stationary phases) chosen. Another operational problem has been minimized by the development of bonded liquid phases which render the stationary phase "more stationary" and eliminate the need to saturate the mobile phase with the stationary phase. Consequently, a wider choice of mobile phases becomes evident and also the latitude of sample type applications. Prior knowledge of the nature of the samples to be analyzed is always necessary. Reversed phase liquid-liquid chromatography has useful applications also. In this type of chromatography, a polar solvent and a non-polar stationary phase are employed. Better success with liquid-liquid chromatography, as with liquid-solid chromatography, is realized by careful choice and pre-testing of the solvent systems.

3. Ion-exchange chromatography, comprised of macro- and micro-reticular resins and pellicular resins (see Figure 1-2) (4), provides separations on the basis of the ionic properties of the compounds to be separated. The process involves reversible exchanges of ions between the ion exchange phase and the components to be separated. The ion exchange process was one of the earliest chromatographic techniques used and has found some of its greatest applications in the biological field.

Synthetic resins (cross-linked polystyrene) are coated or bonded with an organic phase terminated with anionic or cationic functional groups. Typical functional groups are as follows:

Strong cation exchange = $-SO_3^-H^+$

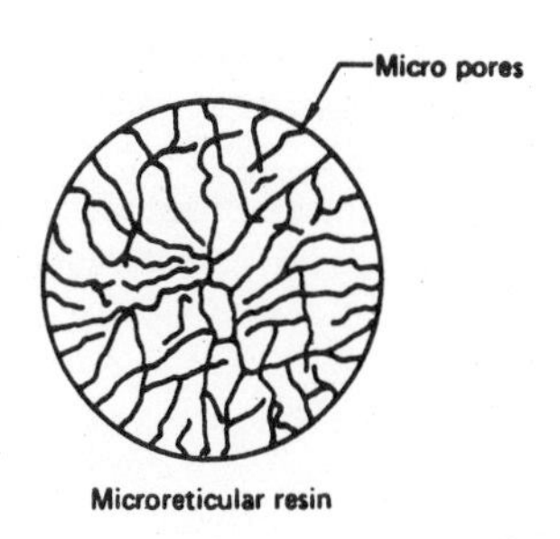

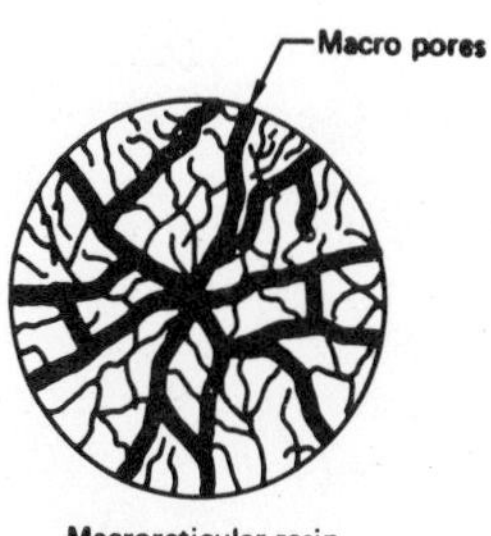

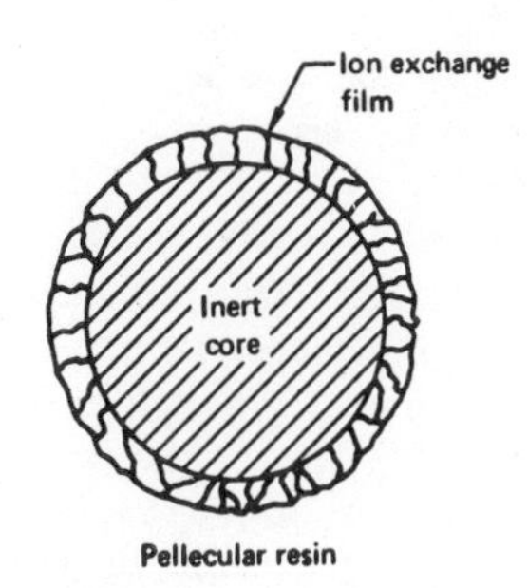

Fig. 1-2. Structural types of ion exchange resins used in liquid chromatography (Courtesy of C. D. Scott, Oak Ridge National Laboratory).

Weak cation exchange = $-COO^{-}Na^{+}$

Strong anion exchange = $-CH_2N^{+}(CH_3)_3Cl^{-}$

Weak anion exchange = $N^{+}H(R)_2Cl^{-}$

The exchange capacity of the weak cation and anion resins is dependent upon pH over a reasonably narrow range; whereas, the strong cation and anion resins have exchange capacity over a much broader range and are, for all practical purposes, independent of pH within these limits.

Aqueous, low concentration buffered solutions are normally used; however, some applications demand mixtures of organic and aqueous solvents. The amount and nature of the organic solvent should be carefully selected, otherwise the coated organic phases may be stripped from the polymer bead and render the column useless for ion exchange applications.

In many cases, the gradient elution technique (described in Chapter 2, Solvent Transport Systems) has drastically increased the efficiency of ion exchange chromatography. For example, many applications in which the components to be separated are extremely sensitive to pH change gradients may be established along the length of the column and achieve separations otherwise impossible.

In the analysis of unknown mixtures, frequently there are components which are irreversibly adsorbed on the resin. Prolonged usage under these conditions generally leads to poor column performance. This may be eliminated by using a short pre-column of the resin and periodically replacing it with a freshly prepared one.

Sample overload is often the cause for lack of resolution, especially when using small internal bore columns packed with pellicular resins. These columns can generally accommodate approximately 50 micrograms of component before overloading becomes evident.

Future developments in ion exchange chromatography will be directed toward increasing the speed of analysis. This will be accomplished, in part, through increased knowledge in the preparation of pellicular resins. Those pellicular resins which are permanently bonded seem to provide columns of higher speed, efficiency, and longer column life in comparison with the "coated" sphere. The major difficulty encountered with these resins is that of manufacturing processes. Wide variations in performance have been observed within different batches of the same resin. High-speed preparative ion exchange chromatography should be a most promising development in the near future. Scale-up technology and the evaluation of resins with new functionality will be critical areas for research.

4. Exclusion (or) Gel Permeation Chromatography is a separation of the components in accordance with their molecular size. The heavier components, exhibiting less permeation into the gel, are eluted from the column first. This is probably the simplest form of chromatography and is widely applied in the field of polymer studies. Data is obtained in the form of molecular weight distributions and for this reason is very useful for surveying unknown sample mixtures. A gel matrix provides the separation of various molecular weight species depending upon the size of the gel particles. Some gels possess active sites, and separations based strictly on molecular size can be drastically altered by affinity or steric effects.

A-plot (Figure 1-3) of molecular size, in Angstrom units, versus elution volume (mls.), shows the exclusion limit of the gel, whereby molecules of this size exhibit no permeation into the gel pores and are totally excluded. The exclusion limit is also defined as the interstitial volume or the volume of liquid in the column between the gel beads and is available for removing from the column those molecules totally excluded by the gel. The total penetration volume is defined as that portion of the column available to smaller molecules. The region of selective permeation is merely the difference between the exclusion limit or

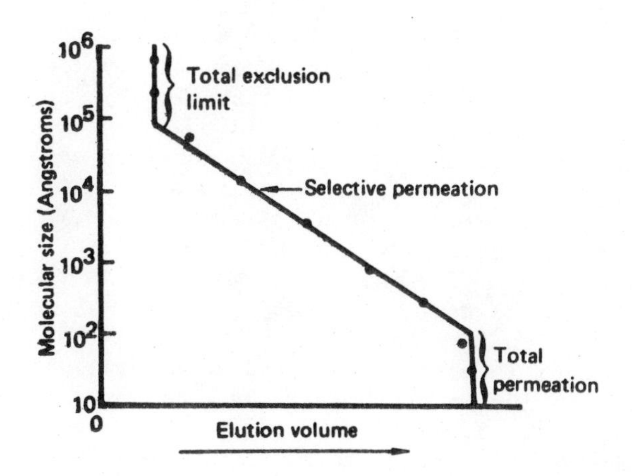

Fig. 1-3. Ideal gel permeation calibration curve showing regions of exclusion, selective permeation, and total permeation obtained from the plot of log molecular weight versus elution volume.

interstitial volume and total permeation and is the selective pore volume of the column providing a plot of the elution volume versus the logarithm of the molecule size.

The first objective in any chromatographic system is that of obtaining adequate separations of the components being sought. Separation is, of course, dependent upon the resolving power (resolution, R_S) provided by the column. In general practice, the ability of a column to separate a given pair of compounds can be measured in terms of resolution. As seen in Figure 7-4, resolution is a function of peak width and distance between peaks and is determined from the relation

$$R = 2\Delta t_R/W_1+W_2 \;\tilde{=}\; \Delta t_R/W_2$$

where R is the column resolution, W_1 and W_2 are peak widths at the baseline and Δt_R is the distance between the apex of the two peaks.

Resolution may also be defined as follows:

$$R_S = 1/4\ (\alpha-1/\alpha)\ (k'/1+k')\ (\sqrt{N})$$

SELECTIVITY, CAPACITY, RANDOM DISPERSION

The relative retention, α, of a column for two components is dependent upon the ratio of the distribution coefficients, k2'/kl' and describes the relative selective retention of the components by the stationary phase. The capacity factor, $k' = V_1 - V_0/V_0$, predominantly governs the length of time required to obtain the desired separation. Wide variations in k' values for multicomponent systems define the "general elution problem". k' is the rate of the amount of solute in the stationary phase versus the amount of solute in the mobile phase. It may be altered by varying the eluant and the type or amount of stationary phase. The random dispersion term, N, relating band spreading and retention volume, is a measure of column efficiency and descibes the fashion in which variables associated with the number of plates, i.e., flow rate, column, eluant, etc., alter the effects of other terms in the resolution equation. This may be determined from the relation

$$N = 16\left(\frac{t_R}{W}\right)^2$$

and is illustrated in Figure 7-3.

Efficiency is a function of column length and columns may be compared by expressing this in terms of a Height Equivalent to a Theoretical Plate or HETP. This may be determined from the relation H = L/N where L is the column length, N is the number of theoretical plates, and H = HETP.

The Van Deemter equation, $H = A + \frac{B}{v} + Cv$, may be used to clarify further the effects of the random dispersion term in the resolution equation. The A term in the Van Deemter equation is the phenomenon of eddy diffusion or flow inhomogeneity. The B term is related strictly to band broadening (5) via the Einstein equation for diffusion. The C term is related to the forces of nonequilibria resulting from the resistance to mass transfer (6). Figure 1-4 is a plot of H versus v for a retained solute in both gas and liquid chromatography, where H is the height equivalent to a theoretical plate (HETP), and v is the mobile phase velocity. Optimum separations are those in which resolution is achieved in minimal time; therefore, the variation of H with v can be seen from:

$$t_R = N(1+k')H/v$$

where t_R (retention time) is directly proportional to H/v considering that values for the N term (theoretical plate count) are contingent on R_s (resolution) and the α (relative retention) value. The curves in Figure 1-4 for GC and

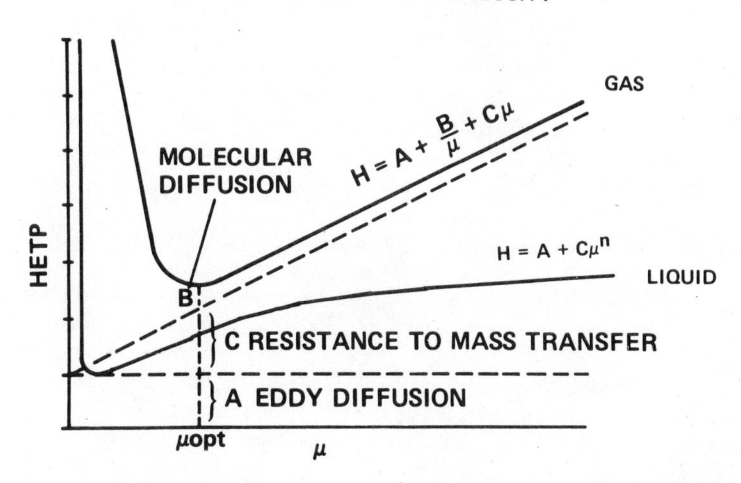

Fig. 1-4. HETP versus mobile phase velocity for GC and LC.

LC are drastically different; and these differences are attributed to the difference in diffusion coefficients between gases and liquids, the latter being 10^4-10^5 times smaller (2). Therefore, in liquid chromatography effects of the B term in the Van Deemter equation are negligible and the equation takes the form H = A + Cv.

The theoretical considerations outlined above can only be optimized on a potentially productive liquid chromatographic system, i.e., the system must be properly maintained, and the operator must be clearly aware of potential hazards and troubleshooting tips to prevent costly downtime. Routine maintenance of the entire system, such as periodic cleaning using solvents of different polarity, is an excellent preventive measure for eliminating noise problems and unnecessary pressure drop. Inspection of all fittings as well as seals and gaskets within the pumping system may eliminate costly downtime and/or pump repair. A daily check with a suitable standard mixture provides a current assessment of column performance and changes in detector sensitivity. If one establishes a preventive maintenance program for his liquid chromatographic system, optimization of the three terms in the resolution equation is enhanced and additional time is realized for more productive pursuits.

GLOSSARY (9)

ADSORPTION: See LIQUID/SOLID.

AFFINITY: The general term for all modes of separation which rely on relative attraction between the surface of the packing material and the components in solution. See also LIQUID/LIQUID, LIQUID/SOLID and ION EXCHANGE.

ANALYTICAL CHROMATOGRAPHY: Operation at any sample loading level in which the objective is a recorder trace representing the components present. Identification is made by comparing the characteristic elution volumes of known standards.

ALPHA (α): A measure of the separation of two components at their peaks. It does not take into account whether or not any substantial part of the peak area has overlapped. See also BASELINE RESOLUTION.

BASELINE RESOLUTION: Separation at the peak base, that is, no overlap of any peak area.

BONDED PHASE PACKINGS: Selected compounds are permanently attached to the packing particle by chemical bonding. Avoids "bleed-off" problems inherent in conventionally coated, Liquid/Liquid chromatography packings.

CAPACITY, k': A measure of the solvent volume required to elute a component from the column, expressed as multiples of the column void volume.

EFFICIENCY: A standard of column performance relating to the amount of peak spreading which occurs. See PLATES.

ELUTION VOLUME: The solvent volume required to elute a species from the column. See CAPACITY.

ELUTROPIC SERIES: A graphical ranking of relative solvent polarities.

EXCLUSION LIMIT: A rating of the size sorting capacity of GPC packings. The value is the nominal diameter of the largest packing pores. Larger molecules are excluded and pass directly through the column while the smaller molecules can enter the packing, thereby delaying their rate of travel through the column.

GEL PERMEATION (GPC): A mode of liquid chromatography in which samples are separated according to size. The packing consists of porous particles with controlled pore sizes. The smallest sample components can enter these pores while the larger components, by virtue of their size, are excluded from the internal regions of the bead and, therefore, travel through the chromatographic column in the interstitial space between the packing articles. Consequently, the smaller components take longer to elute while the largest components elute first.

GRADIENT ELUTION: A technique of varying the composition of the moving solvent in a predetermined manner to insure the elution of all peaks within a reasonable period of time. Generally used when some components in the sample elute within a reasonable time, capacities of 2 to 8, while other components remain on the column much longer. The solvent composition can vary continuously during the course of the run or can be changed as a step increment.

ION EXCHANGE CHROMATOGRAPHY: The mode in which the relative binding equilibria of ionizable components in the sample with ion exchange sites on the packing is the basis for separating components.

LIQUID/SOLID CHROMATOGRAPHY: An affinity mode in which the sample components are separated based on their relative solubilities in the moving liquid phase and a second immiscible phase which is coated on a solid support. A recent advance in liquid-liquid chromatography is to permanently bond the coating to the substrate to avoid certain operational problems. See also BONDED PHASE PACKINGS.

LIQUID/LIQUID CHROMATOGRAPHY: An affinity mode of chromatography in which separation of the sample components is based on differences in adsorption on the surface of solid packing articles. See also REVERSE PHASE and NORMAL PHASE.

NORMAL PHASE CHROMATOGRAPHY: Affinity modes in which the packing surface is quite polar and the moving solvent is relatively non-polar, e.g., chloroform across silica gel columns. See REVERSE PHASE.

PARTITION: See LIQUID/LIQUID.

PELLICULAR PACKING: A recent advance in packing technology consisting of a solid core and a thin outer crust of active porous material. This provides rapid mass transfer between the sample and the bead, leading to rapid, high efficiency separations.

PLATES: A measure of the efficiency of the column which takes into account the elution volume and also the peak width at the baseline. The narrower the peak, the more efficient the column and the higher the plate number will be.

POLARITY: A general characteristic of materials arising from the presence of electro-negative and electro-positive groups in the compound. The number, strength and separation of these groups in the molecule all contribute to polarity. Polarity influences molecular interactions, including solvent-sample, solvent-packing and sample-packing affinities. See ELUOTROPIC SERIES.

POROUS PACKING: Packing materials which are porous throughout the entire structure of the particle.

PREPARATIVE LIQUID CHROMATOGRAPHY: Any scale of operation in which the objective is the collection of sample components for subsequent identification or use. It may involve submilligram or gram quantities of material.

RECYCLE: A technique in which the unresolved or partially resolved sample eluting from the column flow through the detector and then back into the pumping system for as many additional passes through the column as necessary to achieve satisfactory resolution.

RESOLUTION, R: The total measure of component peak separation at their apexes and at their baselines. Components with R = 1 are 98% resolved.

REVERSE PHASE: Affinity modes in which the packing surface is non-polar and the moving solvent is very polar, e.g., water or methanol across BONDAPAK C_{18}/CORASIL, a hydrocarbon modified surface. See BONDED PHASE.

SIZE SEPARATION: See GEL PERMEATION.

SOLVENT PROGRAMMING: See GRADIENT.

VOID VOLUME: The total unoccupied volume in a packed column, consisting of the volume between the packing particles (the interstitial volume) and the porosity of the packing material. Void volumes are typically 40 to 80% of the column and are determined by injecting a non-retained component.

REFERENCES

1. L. R. Snyder, *Principles of Adsorption Chromatography*, Marcel-Dekker, Inc., New York, 1968.

2. J. J. Kirkland, Ed., *Modern Practice of Liquid Chromatography*, John Wiley and Sons, Inc., 1971

3. T. N. Tischer and A. D. Baitsholts, "Thin-layer Chromatography: Why and How," *American Laboratory*, p. 72, May 1970.

4. C. D. Scott, "Ion Exchange Chromatography" Lecture Outline for Liquid Chromatography Course, Washington University, St. Louis, Missouri, 1970.

5. A. Einstein, Z. *Elektrochem., 14,* 1908.

6. J. C. Giddings, *Dynamics of Chromatography*, Marcel-Dekker, New York, 1965, pp. 190-193.

7. *Basic Liquid Chromatography*, Varian Aerograph, Walnut Creek, CA., 1971.

8. L. R. Snyder and J. J. Kirkland, "*Introduction to Modern Liquid Chromatography*," Wiley-Interscience, 1974.

9. Waters Associates Inc., "A Liquid Chromatography Glossary," DS 011, December 1973.

QUESTIONS AND ANSWERS:

Question: Why is the elution order of components in thin layer chromatography sometimes different than those obtained in high speed liquid-solid or adsorption chromatography?

Answer: There are several reasons why this difference may exist and perhaps, one of the most significant is the effect of the binder material associated with the TLC plate. Also, different solvent combinations used in the two techniques will produce different levels of activity and selectivity of the silica plate compared to the silica packed column. Snyder has recommended pre-saturation of all solvents with water prior to use in high speed LSC to maintain more reproducible system with regard to the degree of surface activation.

Question: What are the major causes of tailing and memory effects in adsorption chromatography?

Answer: This may be due primarily to insufficient deactivation of the silica or alumina surface in the column. In these cases, prior saturation of the mobile phase with water or the addition of small amounts of methanol to the mobile phase should provide a reproducible and satisfactory level of deactivation.

Another common cause of both tailing and memory effects or "ghosting" is that of the sample injector. The injection system should be designed such that it is cleanly swept with the mobile phase and should also introduce minimal dead volume to the system.

Question: In liquid-liquid chromatography, what causes increasing peak tailing and broadening on consecutive runs on the same sample during the course of a day?

Answer: The mobile phase being used is "stripping" the coated stationary phase from the support. The recommended procedure is to either use a pre-column containing an excessive loading (∿30%) of the stationary phase or pre-saturate the mobile phase with the stationary phase such that the level of coating is not depleted. More recently, permanently bonded stationary phases have become available, eliminating the need for pre-columns and pre-saturation.

Question: In exclusion chromatography, specifically gel permeation on smaller molecules, i.e., less than mol. wt. 1000, why does one obtain different calibration curves for different classes of compounds even when the same column system and solvent are used?

Answer: The answer to this question is at least two-fold. First of all, different solvents will cause the gel matrix to behave differently with regard to expansion and/or contraction which can alter the pore size of the gel. Secondly, the geometry of the molecules to be separated do have a pronounced effect as well. For example, consider two different classes of compounds having approximately the same molecular weight. We can select oligomers of polystyrene and polypropylene glycols. Obviously, the geometries of these two classes of compounds are different and consequently do not exhibit the same degree of permeation into a gel pore matrix; therefore, the resulting calibration curves are different. Another phenomenon which can alter predicted results is that of active sights in the gel matrix which may produce some affinity considerations, as well as size or exclusion principles.

Question: After being involved in the use of gas chromatography for several years, don't you feel the recent publicity and importance given to liquid chromatography is slightly overrated?

Answer: When one considers that, of all the known organic compounds, approximately 25 percent of these can be analyzed by GC and the remaining 75 percent can be analyzed by LC, then it behooves all that are chromatographers to stay abreast of these current developments.

Chapter 2

SOLVENT DELIVERY SYSTEMS

A major requirement of modern-day liquid chromatography is that the solvent delivery system efficiently accommodate the desired solvent and deliver it to the various parts of the system. This capability must exist over a wide range of flow rates and inlet pressures and with all useful solvents. Systems of this type consist of a degasser (for removing dissolved air and other gases), solvent reservoirs, and high-pressure pumps. The early manufacturers of pumps for liquid chromatography (LC) were concerned only with the delivery of mobile phase at a fairly reproducible rate when operating at system back-pressure in the range of 100-500 psi. As sampling systems, column technology, and detectors for LC were improved, the need for improving solvent delivery systems became urgent. General requirements which must be met in order for a pump to be compatible with the other components in a high speed liquid chromatograph are (a) operating pressure to several thousand psi; (b) precise solvent delivery over a relatively broad flow range; and (c) compatibility with a wide choice of solvents. More specific parameters for various modes of operation are outlined in Table 2-1.

In a modern LC pump, the electronics, mechanical linkages, check valves, seals, plungers, cams and fittings generally have been optimized for maximum performance; however, there are some precautionary measures the chromatographer

TABLE 2-1 (1)

Recommended pump parameters

Parameter	Mode of Operation		
	Research	Quality Control	Preparative
Pressure capability (psi)	5000	3000	500 - 1500
Flow (a) range (ml. per min.)	5	5	10 - 20
(b) accuracy	±5%	±5%	±10%
(c) reproducibility	±2%	±2%	± 5%
Solvent Storage	unlimited	200 - 500 ml.	unlimited
Gradient elution capability	necessary	not critical	not critical
Pulse-free delivery	necessary	necessary	not critical

must cleatly recognize as his own responsibility, regardless of the type pump chosen. For example, many unwarranted complaints regarding pumping systems occur because the chromatographer uses corrosive solvents or fails to filter and adequately degas the mobile phase.

Automatic solvent degassing systems offer time savings and more efficient operation. All aqueous solvent systems and many organic systems should be degassed. Manual degassing of solvents is effective but quite time consuming. Manual degassing of aqueous solvents may be accomplished by placing the solvent on a hot plate and heating to a slow boil. The boiled solvent is then cooled rapidly. Solvents may also be adequately degassed under vacuum with gentle stirring for five to ten minutes. Generally, the use of stored degassed solvents is not recommended because, on standing, sufficient air is re-dissolved to render the degassing treatment ineffective. Dissolved gases in solvents have a tendency to produce bubbles which may become localized in any part of the system having sufficient dead-volume. Possible dead-volume areas are the sample injection system, column and detector. Some commercial degassing systems are of the controlled vacuum type and are quite efficient. Many of these provide auxiliary storage chambers designed for short-term storage, i.e., for preparative work, when larger quantities of solvent are frequently needed. If a conventional forepump or oil diffusion pump is used as the controlled vacuum source, periodic checks regarding the oil level and condition of the oil in the pump should be made.

Solvent change-over may cause many problems, most of which may be almost entirely eliminated by exercising a few precautions within the solvent transport system. Using miscible solvents and thorough cleaning of the reservoirs (for degassing and solvent storage), the pumping system, column, and detector before introducing the desired solvent are worthwhile. Many solvents off the shelf contain particulate matter which may be removed by filtration. Filtering of all solvents is recommended before degassing.

Efficient, high-pressure pumps are required to accomplish "high speed" liquid chromatography. These pumps must be of durable construction and all materials of construction must be as inert and corrosion resistant as possible. Metal parts are generally made of #304 and/or #316 stainless steel, the latter being preferable. The gaskets and seals are usually made of Teflon or Kel F. Many times solvent leakage may occur around check valves and purge valve seals. This is often due to physical damage or foreign matter in the solvent. If damaged, these valves may be easily replaced by the user. A spare parts kit is available from the manufacturer for practically all the pumping systems commercially available. It behooves one to have this on hand in case of pump failure. However, one should not be too eager to replace these components until other possible causes have been eliminated. Some possible symptoms and causes are listed in Table 2-2.

Solvent delivery systems may be generally described as being of two basic types: *constant pressure* or *constant volume*. Both usually operate by displacement of liquid from a chamber by a hydraulic piston. The major difference is that the piston of a constant pressure pump is operated by gas pressure supplied from an external source, while the piston of a constant volume pump is operated by a mechanical linkage. Common exceptions to this description are gas displacement and diaphragm pumps. Comparisons of pumps may also be made with respect to the size of their hydraulic chambers; i.e., that part of the pump containing the volume of pressurized solvent to be delivered before refilling. Small volume reciprocating piston pumps may have chamber volumes of only a few hundred microliters while the screw-driven syringe types may have chamber volumes of several hundred milliliters. Each has advantages and disadvantages. For purposes of comparison, pumps will be discussed in the order of basic type, i.e., constant pressure and constant volume.

The pneumatic amplifier pump, shown in Figure 2-1 (1), is a constant pressure pump and is available with both large and small hydraulic chamber volumes. In this type pump, an external gas supply (air or nitrogen) is delivered to a gas piston of large surface area attached to a hydraulic piston of small surface area. Amplification is usually about 50 to 1 to that inlet pressure capability to approximately 5000 psig is possible with 100 psi gas supply. The hydraulic chamber may be connected to a reservoir through a check valve. In automatic versions, the gas pressure is reversed at the end of the pump stroke and the pump refills instantly. A check valve between the pump and the column prevents flow reversal during the refill stroke. In this type system, performance is again dependent upon properly functioning seals and check valves. Significant flow interruption can occur during refill if the column check valve does not function perfectly. Flow variations can also occur when gas supply pressure varies or when restrictions develop within the system. Restrictions may be caused when particles of septa, insoluble sample constituents or matter from the mobile phase accumulate in the injector or on the head of the column.

Another type of constant pressure system is the gas displacement pump in which a continuous coil of stainless steel tubing forms the hydraulic chamber. The volume of solvent required to fill the coil may vary over the range

TABLE 2-2

Causes associated with pump malfunctions

Symptom	Possible cause and remedy
1. Pump fails to buildup pressure	1. Solvent reservoir empty 2. Improperly closed valve(s). 3. Defective solvent line or connection. 4. Air in pump, may require priming. 5. Oil level too low for maximum efficiency. 6. Leaking seals, check valves and gaskets (refer to specific pump manual for replacement).
2. Pump pressure high enough but no flow through column or detector.	1. Pressure buildup due to excessive pressure drop. Check for plugging in transfer line to injector, column, and detector. Flow is probably escaping through relief valve. 2. Leak in system. Check all fittings, septa, etc.
3. Noisy, erratic and/or pulsating recorder trace.	1. Pulsed pumps - under-damped. Should incorporate additional pulse-dampener. 2. Air bubbles entering detector. Solvents and/or column needs additional degassing or purging. 3. Solid, particulate matter from column entering detector. Detach column and backflush detector and transfer line from column to detector.

TABLE 2-2 - Cont'd.

Causes associated with pump malfunctions

Symptom	Possible cause and remedy
3. - Cont'd.	4. Previous solvent not completely removed from system. Use step-wise miscible solvent purging if consecutive solvents are immiscible. 5. Detector leaking eluant. Remove covers and inspect for leaking seals or damaged cells. 6. Recorder problems: a. Dirty slide wire - clean with aerosol slide wire cleaner and/or methanol. b. Weak tubes in amplifier circuit. c. Mechanical binding of pen and/or slide wire. d. Inadequate connection of signal leads to recorder. e. Grounding problem. f. Gain adjustment too high.
4. During operation, pressure in pump: (a) Increases	1. Increases in pressure during operation is indicative of plugging somewhere in the system. Suspect locations are pulse dampeners, sample injectors (especially if septum injection is being used), column, detector.
(b) Decreases	2. Decreases in pressure during operation is positive indication of a leak in the system. Check points are septa, seals in loop injectors, all fittings, cell (cell windows or spacers may be defective).

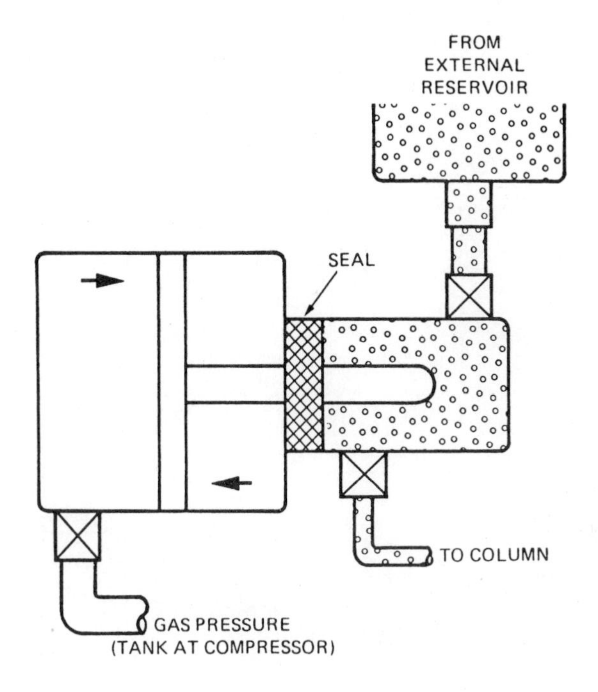

Fig. 2-1. Pneumatic Amplifier Pump.

200 to 500 mls. The coil is filled by gravity through a valve connected to an external reservoir. During operation, the solvent contained in the coil is displaced by gas pressure, usually nitrogen. Figure 2-2 (1) is a block diagram of a typical gas displacement pump in which the valves are interlocked to prevent accidental discharge of gas pressure through the filling reservoir. Systems of this type are rugged, reliable and economically attractive. In addition, mobile phase changes can be made rapidly and without wasting large quantities of solvent. Inlet pressures to 3000 psi can be attained; and, because solvent delivery is from a constant pressure source, the system is inherently pulse-free. During operation, the gas in contact with the liquid in the coil will diffuse into the liquid and become dissolved in it. Because of the small inside diameter and long length of the coil only a small portion at the gas-liquid interface is affected during a normal operating period. This portion is removed and discarded in the next solvent filling operation. The gas displacement pump has limited delivery volume and is subject to the same disadvantages as other constant pressure systems, i.e., flow sensitivity to variations in gas pressure and restrictions in the solvent stream.

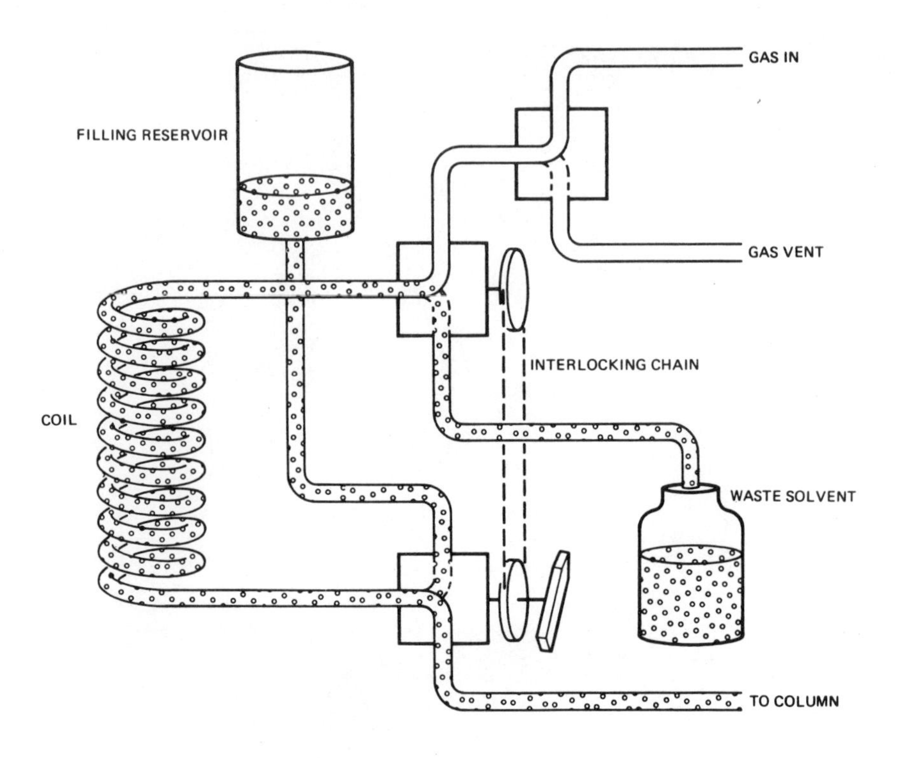

Fig. 2-2. Typical gas displacement pump.

Figure 2-3 (1) is an example of a constant volume and large volume screw-driven piston pump. These pumps are inherently pulse-free and accurate as long as seal integrity is maintained. On the other hand, they are limited in flow rate range and flow must be interrupted during refill. Solvent delivery is accomplished using a variable speed motor to turn a screw which drives a hydraulic piston. Motor malfunction or leaking seals and valves are the most frequent causes for inadequate solvent delivery. When it becomes necessary to refill the pump, the screw-action must be reversed to withdraw the plunger. This may require several minutes for complete refill and may take considerably longer when rinsing of the hydraulic chamber is necessary. In one design, the reservoir may be physically removed and manually refilled.

A constant volume reciprocating piston pump is shown in Figure 2-4 (1). This type pump consists of a small piston attached to a motor via gears and a cam which drives it back and forth in a hydraulic chamber. Pistons are typically fabricated from borosilicate glass, sapphire or chrome-plated

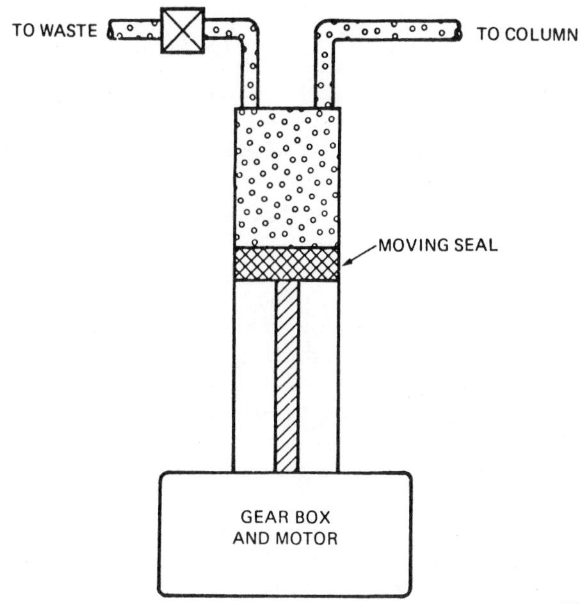

Fig. 2-3. Large volume screw-driven piston pump.[1]

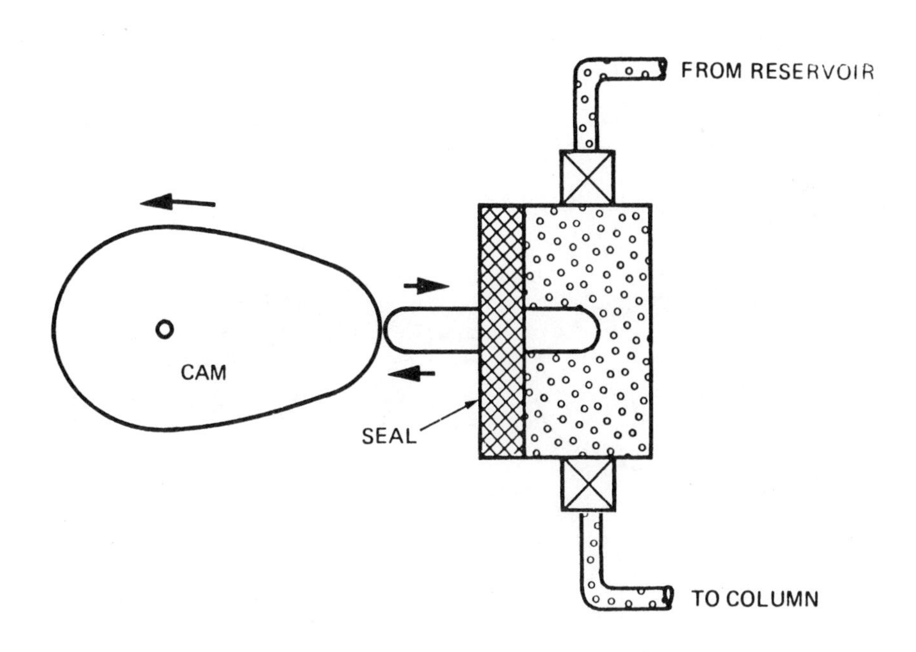

Fig. 2-4. Constant volume reciprocating piston pump.[1]

stainless steel. During assembly or maintenance, alignment of these pistons is critical. Check valves at the inlet and outlet of the hydraulic chamber are also an essential part of the pump design and must function perfectly in order for the pump to deliver accurate flow of solvent. Principle advantages of this type pump are that it draws solvent from an external reservoir and delivers a truly continuous flow. In addition, the small volume reciprocating pumps may be rapidly changed from one mobile phase to another with virtually no waste of solvent. A disadvantage of small volume single piston pumps has been a phenomenon known as "pump-noise," caused by flow pulsations through the column and detector. Various attempts have been made to minimize the noise by incorporating several types of pulse dampeners; however, the pulsations still exist at high detector sensitivities, and other problems may be introduced by the damping device. Pulse dampeners of the small volume, elastic type are easily damaged and may severely limit the maximum operating pressure of the pumping system. Larger column dampeners which are less elastic permit the pump to operate at higher pressures but require excessive time for purging the system during solvent changeover.

The diaphragm pump shown in Figure 2-5 (1) is a special case of a piston pump where the piston is mechanically

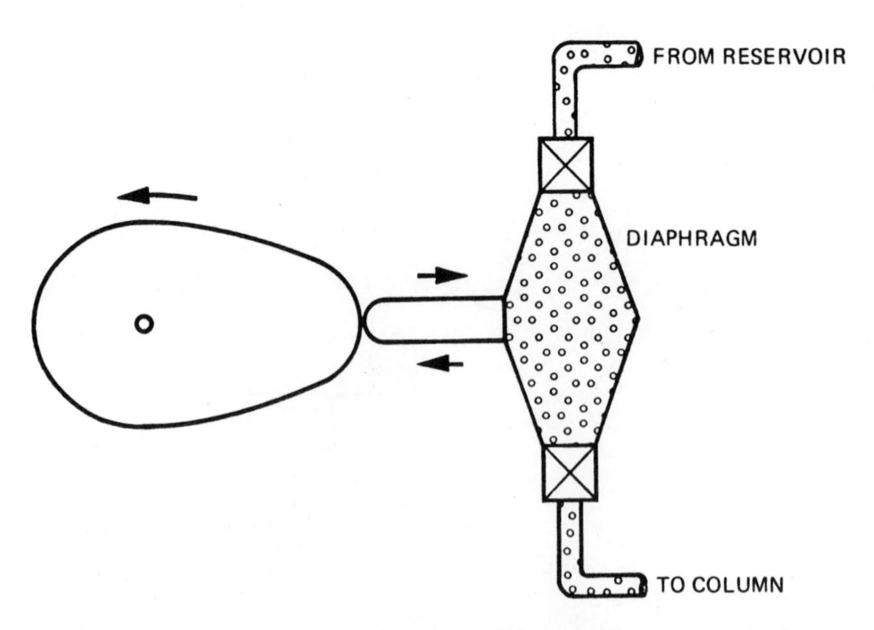

Fig. 2-5. Reciprocating diaphragm pump.[1]

attached or linked hydraulically to the diaphragm, thus avoiding contact of the piston with the mobile phase. In this manner, diaphragm pumps eliminate the need for a high pressure liquid seal around the piston. Diaphragm pumps are, however, subject to problems of pulsating delivery and decreasing flow rate with increasing backpressure. This latter phenomenon is due to the elasticity of the diaphragm which takes up increasing amounts of the piston displacement volume at high back pressures.

A more advanced design in reciprocating pumps is seen in Figure 2-6 (1) and consists of dual reciprocating pistons operating almost 180° out of phase so that the pulse of flow interruption created by the refilling of a single piston is largely cancelled by the action of a second piston operating in an opposing direction. This dual piston design is an improvement over the single piston type and permits high sensitivity monitoring with ultra-violet and refractive index detectors.

Another development in pumps of the dual reciprocating piston type is shown schematically in Figure 2-7 (1). This pump features closed-loop, flow feedback control designed to eliminate both flow irregularities and pump pulsations regardless of system back pressure. An important part of this pump is a device to measure continuous flow. When flow deviations are detected they are compensated by feedback to a pump circuit. As in other designs, properly functioning seals and check valves govern the performance of these systems.

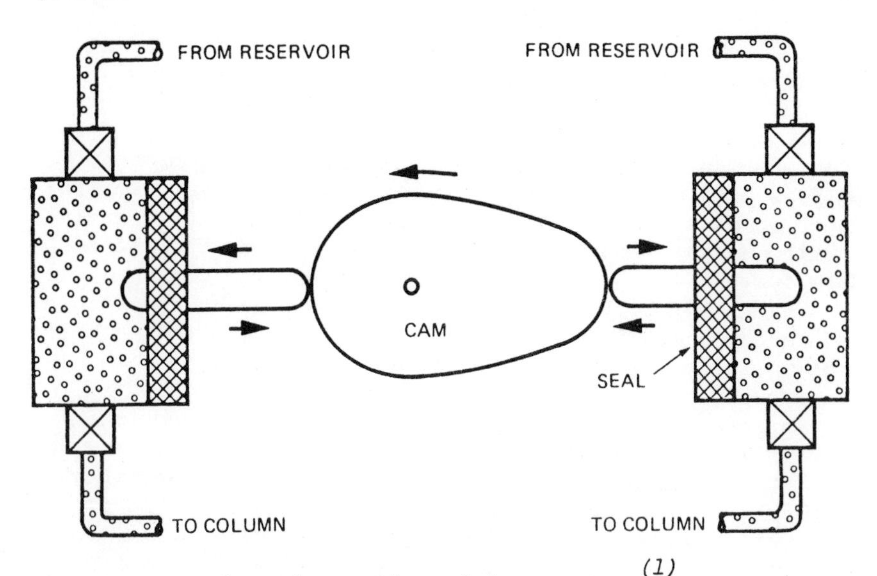

Fig. 2-6. Dual reciprocating piston pump. (1)

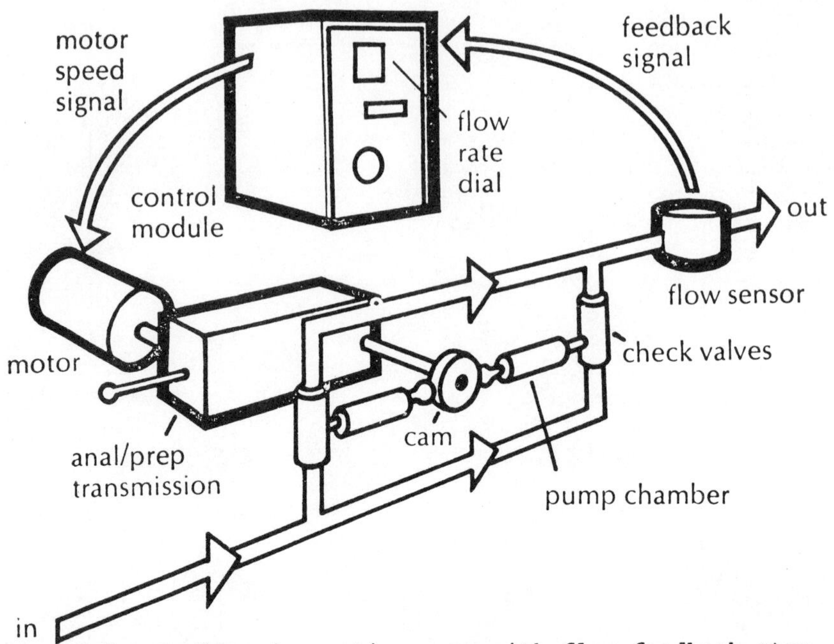

Fig. 2-7. Dual reciprocating pump with flow feedback cton control.[1] *(Courtesy of Spectra Physics)*

GRADIENT ELUTION

The advantages of gradient elution in liquid chromatography are analogous to those of temperature programming in gas chromatography. To the liquid chromatographer, gradient elution provides a reasonable solution to the "general elution problem," i.e., analysis of samples whose components possess a wide range of distribution coefficients. It provides a means for quickly surveying unknown mixtures and permits rapid evaluation of solvent systems during methods development.

The use of Table 2-3 (2) will prove to be an invaluable reference in all forms of LC, especially in gradient elution where all of the listed solvent parameters must be considered.

An ideal gradient system should provide reproducibility, versatility, rapid equilibration and ease of operation. The reproducibility of the system is most important, and the primary factor in determining reproducibility is that of pump performance. The gradient elution system should also be versatile and have the capability of generating concave and convex and linear gradients. Varying degrees of

TABLE 2-3 (2) PROPERTIES OF CHROMATOGRAPHIC SOLVENTS

SOLVENT	$\varepsilon^0(Al_2O_3)$	Viscosity (cP,20°	RI	UV cutoff nm
Flouroalkanes	-0.25		1.25	
n-Pentane	0.00	0.23	1.358	210
Isooctane	0.01	0.54	1.404	210
Petroleum ether Skellysolve B, etc.	0.01	0.3		210
n-Decane	0.04	0.92	1.412	210
Cyclohexane	0.04	1.00	1.427	210
Cyclopentane	0.05	0.47	1.406	210
Diisobutylene	0.06		1.411	
1-Pentene	0.08		1.371	
Carbon disulfide	0.15	0.37	1.626	380
Carbon tetrachloride	0.18	0.97	1.466	265
Amyl chloride	0.26	0.43	1.413	225
Xylene	0.26	0.62-0.81	1.50	290
i-Propyl ether	0.28	0.37	1.368	220
i-Propyl chloride	0.29	0.33	1.378	225
Toluene	0.29	0.59	1.496	285
n-Propyl chloride	0.30	0.35	1.389	225
Chlorobenzene	0.30	0.80	1.525	
Benzene	0.32	0.65	1.501	280
Ethyl bromide	0.37	0.41	1.424	
Ethyl ether	0.38	0.23	1.353	220
Ethyl sulfide	0.38	0.45	1.442	290
Chloroform	0.40	0.57	1.443	245

TABLE 2-3 (2) Continued.

SOLVENT	0(Al_2O_3)	Viscosity (cP,20°)	RI	UV cutoff nm
Methylene chloride	0.42	0.44	1.424	245
Methyl-i-butylketone	0.43		1.394	330
Tetrahydrofurane	0.45		1.408	220
Ethylene dichloride	0.49	0.79	1.445	230
Methylethylketone	0.51	0.43	1.381	230
1-Nitropropane	0.53		1.400	380
Acetone	0.56	0.32	1.359	330
Dioxane	0.56	1.54	1.422	220
Ethyl acetate	0.58	0.45	1.370	260
Methyl acetate	0.60	0.37	1.362	260
Amyl alcohol	0.61	4.1	1.410	210
Dimethyl sulfoxide	0.62	2.24	1.447	
Aniline	0.62	4.4	1.586	
Diethyl amine	0.63	0.38	1.387	275
Nitromethane	0.64	0.67	1.394	380
Acetonitrile	0.65	0.37	1.344	210
Pyridine	0.71	0.94	1.510	305
Butyl cellusolve	0.74			220
i-propanol, n-propanol	0.82	2.3	1.38	210
Ethanol	0.88	1.20	1.361	210
Methanol	0.95	0.60	1.329	210
Ethylene glycol	1.11	19.9	1.427	210
Acetic acid	Large	1.26	1.372	
Water	Very large	1.00	1.333	

curvature in the concave and convex modes are also useful. Rapid equilibration or "turn-around" time between successive runs is extremely important. All features of the gradient system should be easily accessible and electronically controlled such that the operator merely changes switch settings to accomplish his needs. Solvent supply to the system should be adequate for extended analyses.

There are two basic approaches for producing gradients in liquid chromatography. The first approach requires that the gradients be generated in an external reservoir at ambient pressure and then be drawn through a high pressure pump to the column. In this design, the pump external reservoir is the mixing chamber for the gradient solvents and the pump hydraulic chamber must be of the small volume type. Advantages of an external gradient system are lower cost and virtually infinite flexibility with respect to the number of solvents and the manner in which they may be mixed. Unfortunately, external gradients are very time consuming and inconvenient to use because reservoirs must be cleaned and filled with fresh solvent for every gradient. Although external gradients are usually less expensive, it should also be pointed out that the use of this approach does not necessarily eliminate the need for additional pumps to feed gradient solvents into the mixing reservoir.

A second and much more convenient type of gradient system employs two pumps which flow into a high pressure mixing chamber prior to going through the column. Advantages of this type system are rapid turn-around time, ability to generate solvent mixtures of constant composition as well as gradients, plus the amenability to automation. Disadvantages are high cost due to the addition of a second high pressure pump and programmer, and the fact that gradient shapes and durations are usually limited by the design of the programmer. The higher cost of a two-pump gradient system is offset to some extent by the fact that the pumps may be used independently to perform separate solvent delivery operations. In the two-pump gradient elution system, the mixing chamber should be small volume (0.5 to 2.0ml) and should be constructed of material capable of withstanding high pressures. It should be designed so that it is cleanly swept and insures adequate mixing of the solvents. Due to differences in solvent viscosities, physical agitation is usually necessary to insure adequate mixing over a wide range of flow rates. This is commonly accomplished with small magnetic stirrers. A mixing chamber will cause a step-change in solvent composition to be rounded off or smoothed. The response of the output composition to a step-change is exponential with a time constant equal to the internal mixing volume divided by the flow rate. For example, if the internal mixing volume is 1.2 ml and the flow rate is 2 ml/minute, the time constant is 0.6 minute. It would require approximately 1.8 minutes at this flow rate to flush out the chamber contents from one solvent to 95% pure new solvent.

Most of the commercially available gradient systems will perform the desired task of solvent delivery and reproducibility. Most importantly, the operator must define his particular system with regard to dead volume, system anomalies and equilibration time between successive analyses. One good method of checking the precision, accuracy and smoothness of a gradient profile is to place the same solvent in both A and B reservoirs (in the case of a two-pump system) and add a small percentage of a UV absorbing compound to the B reservoir. Gradient profiles can then be recorded using a dummy column or a column filled with inert glass beads connected to a conventional UV detector. Figure 2-8 (1) shows a typical profile for a water to water plus 0.1% (v/v) acetone gradient. Special attention should be given to smoothness and reproducibility, especially at the beginning and end of the gradient run where one pump is phasing in or out at very low flow rates.

Basic requirements for an LC pumping system have been cited to have high pressure capability and solvent compatibility. The ideal pumping system for LC should also provide accurate, precise, and pulse-free solvent delivery over a wide range of flow rates. This becomes true when the data is to be used quantitatively. Although all parts of the LC system are capable of affecting the reliability of quantitation in LC, certainly the pumping system may be considered of utmost importance toward achieving this end. For example, and for purposes of illustration, consider the analysis of a simple three component fused-ring aromatic mixture as shown in Figure 2-9 (1). Table 2-4 (3) shows the effects of fairly wide flow rate areas of the components studied. On an absolute area basis, the percent standard deviation for any given component is 11 to 12 percent when the column temperature is held constant at either 58°C or 29°C. The normalized areas yield percent standard deviations of 0.4 to 0.5 percent at the same temperatures. Most LC pumps are capable of much better accuracy and precision than ± 10 percent; however, this data does indicate the need for maintaining constant flow throughout the system, especially if absolute areas are being used with no internal standardization. Table 2-5 (3) shows the results obtained when both flow rate and temperature are held constant. The percent standard deviation for the absolute area at 58°C and 29°C is 2.3 percent and 1.1 percent, respectively. Comparing data from Tables 2-4 and 2-5 further substantiates the need for optimum flow control.

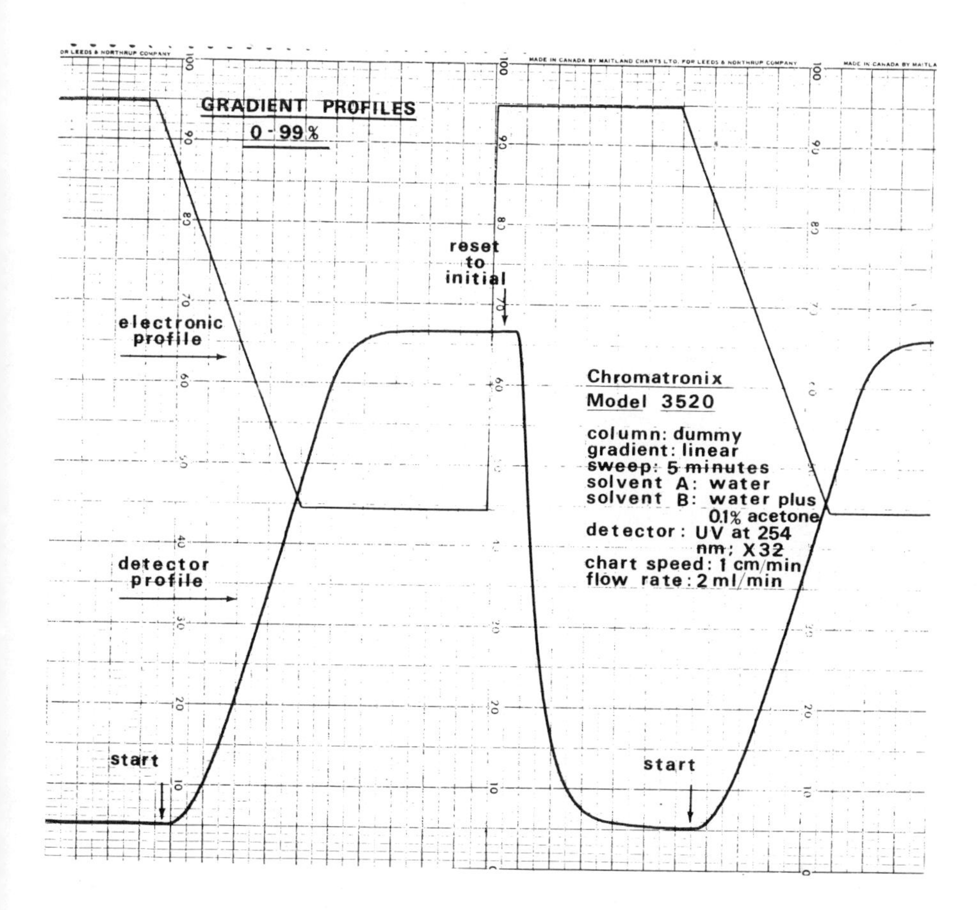

Fig. 2-8. Gradient check-out procedure for water to water + 0.1% acetone.

Cleaning or washing the pumping system is always necessary when the operator desires to change eluants, especially if the eluants have different polarities. A miscible solvent series used in a step-wise manner insures thorough cleaning of the pump. Adequate purging with the ultimate solvent of choice is mandatory. All valves, fittings and transfer lines associated with the pump should likewise be cleaned. Solvents known to be detrimental to proper pump function are strong acids, especially halogen acids, with pH < 2.0 and extremely strong bases. Even short-term use of the above solvents may rapidly cause corrosion and undesirable deposits in valves, valve seats and virtually any metal surface in the pumping system.

In conclusion, virtually any part of the solvent transport system can be rapidly maintained by the operator provided an adequate supply of fittings, valves and seats are conveniently available. The time and dollar investment for replacement spare parts is extremely economical compared to the downtime involved when these components are not on hand.

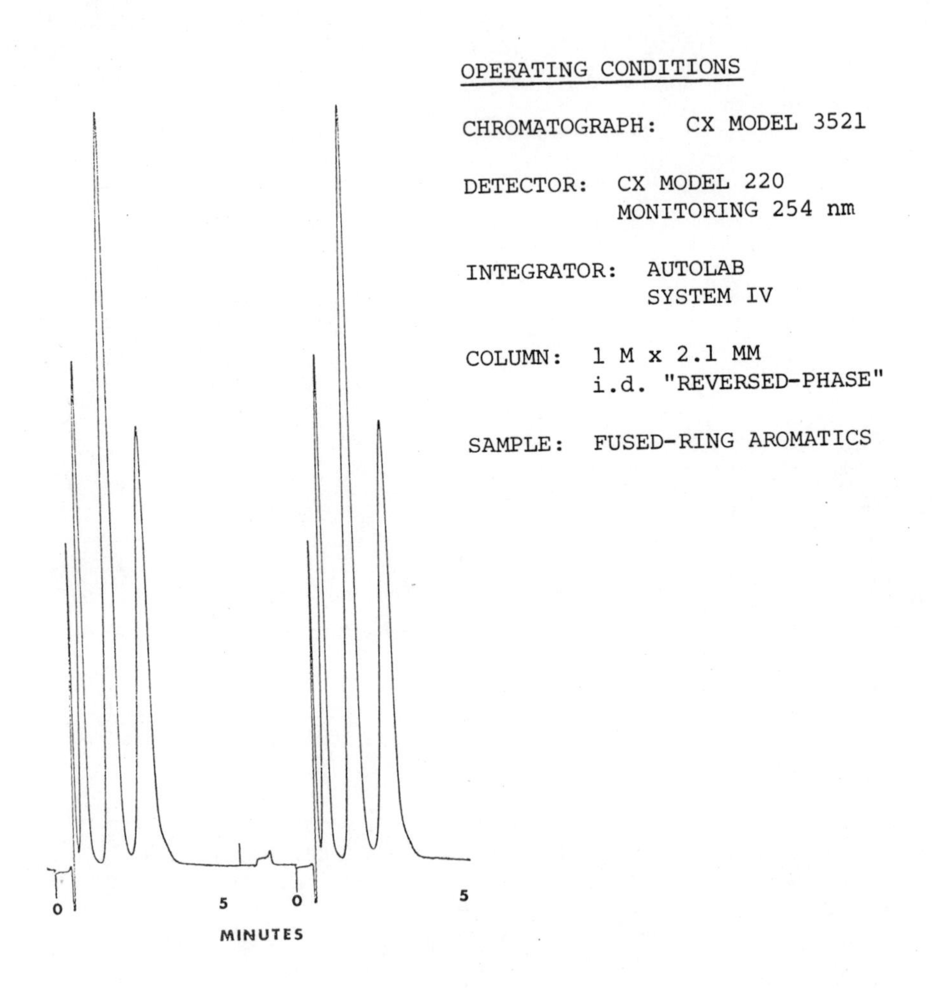

Fig. 2-9. Three component fused ring aromatic mixture used to study flow rate deviations.[3]

TABLE 2-4

Effect of temperature and flow rate on peak areas (variable *flow rate,* constant *temperature)* [3]

Absolute Areas	
σ, 58°C	σ, 29°C
11.1%	12.0%
Normalized Areas	
σ, 58°C	σ, 29°C
0.5%	0.4%

TABLE 2-5

Effect of Temperature and flow rate on peak areas (at constant *flow and* constant *temperature)* [3]

Absolute Areas	
σ, 58°C	σ, 29°C
2.3%	1.1%
Normalized Areas	
σ, 59°C	σ, 29°C
0.4%	0.2%

REFERENCES

1. M. T. Jackson and R. A. Henry *Analytical Instrumentation,* Vol. 12 ISA AID 74424 Instrument Society of America, 20th Annual Meeting, May, 1974.

2. L. R. Snyder, *Principles of Adsorption Chromatography,* Marcel Dekker, Inc., New York, 1968.

3. R. A. Henry, M. T. Jackson, S. Bakalyar, *Quantitative Liquid Chromatography,* Presented at the 1974 Pittsburg Conference.

QUESTIONS AND ANSWERS

Question: Why all the controversy over constant volume and constant pressure pumps when either will do the job?

Answer: Perhaps the greatest controversy has been among manufacturers of the various types. Each type has its own advantages and disadvantages. The choice of pump really is to be governed by its use and the ease of optimum operation.

Question: Most of the current pump designs in liquid chromatography incorporate the use of check valves in some part of the system. What can the user of this equipment do to maintain optimum performance?

Answer: The principle responsibility of the user is that of using "clean" solvents, i.e., free of particulate matter and adequate pump priming. Quantitative filter apparatus may be obtained from Millipore Corporation and has been used with good success. The Figure 2-10 is a drawing of a typical check valve. As can be seen, particulate matter associated with the ball and ball seat can cause inadequate performance.

Question: Are there any problems associated with the use of aqueous buffered solutions in modern LC pumps?

Answer: While operating, generally not. However, it is not recommended that salt solutions be left in the pump when idle due to possible corrosive action. In general, halide salts and halogen acids should be deleted from the user's list entirely, if possible. Aqueous solutions with pH<2 or pH>8-9 should also be deleted from the list.

Question: How important is recycle capability and what are the trade-offs one makes in using this technique?

Answer: Recycle capability, defined as multiple cycling of partially resolved components through the same column, has more than limited use for HSLC. Sometimes, however, the same or better resolution may be obtained by improving the column and mobile phase choices initially.

There are instances where recycle can be used to some advantage. There are at least two ways in which this can be done. Both the "closed-loop" approach described by Bombaugh, Dark and Levangie, *J. of Chromatographic Science*, 7, 42 (1969) and

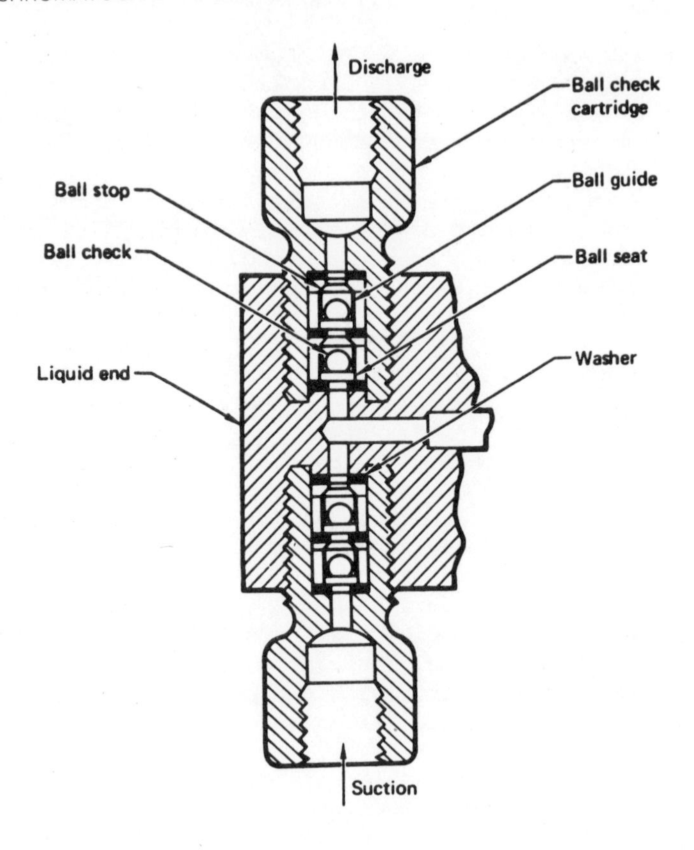

Fig. 2-10. Typical check valve assembly.

the "Alternate Pumping Principle," described by Henry, Byrne, and Hudson, *J. of Chromatographic Science*, *12*, 197 (1974) accomplish the desired task; however, some band broadening occurs in both cases due to the small added dead-volumes in the systems studies.

Question: Are gradient elution techniques capable of generating data for quantitative analyses?

Answer: Yes. Many of the currently available gradient systems may be used quantitatively provided the user is thoroughly familiar with his system. This entails a definition of system dead-volume, reproducibility, turn-around time between analyses, and also chemical knowledge of the system being analyzed. In any event, the use of an internal standard is recommended. The choice of the internal standard and the amount should be closely related to the components being sought in the analysis.

Chapter 3

LC SAMPLE INTRODUCTION SYSTEMS

The sample introduction system in liquid chromatography enables the operator to introduce his sample efficiently into the system without disrupting the established flow equilibrium in the column and detector. This criterion may seem easy to achieve, but only a few of the current designs adequately meet this requirement.

The basic types of sample introduction systems are septum injectors, stopped-flow injectors, and loop injectors. Each of these will be described together with inherent advantages and disadvantages. Ideally, the system of choice should introduce only minimal dead-volume to the system; otherwise, column efficiency will be impaired.

Septum injectors, used most widely in low-pressure systems, permit the operator to introduce his sample, virtually on-column. As in gas chromatography, this seems to be an extremely efficient technique for maintaining and optimizing the theoretical plate-count of the column. Septum injectors, of proper design (Figure 3-1) (1), do provide capabilities for minimizing dead-volume in the system. Normally, 5-50 μl syringes are used for introducing the sample. This technique is quite useful in the 100-1000 psi range of pressure being delivered; however, at pressures above 1000 psi problems become evident. Possibly, the major limitation is the choice of septum materials to withstand the high pressures in the system and to be compatible with a wide range of solvents.

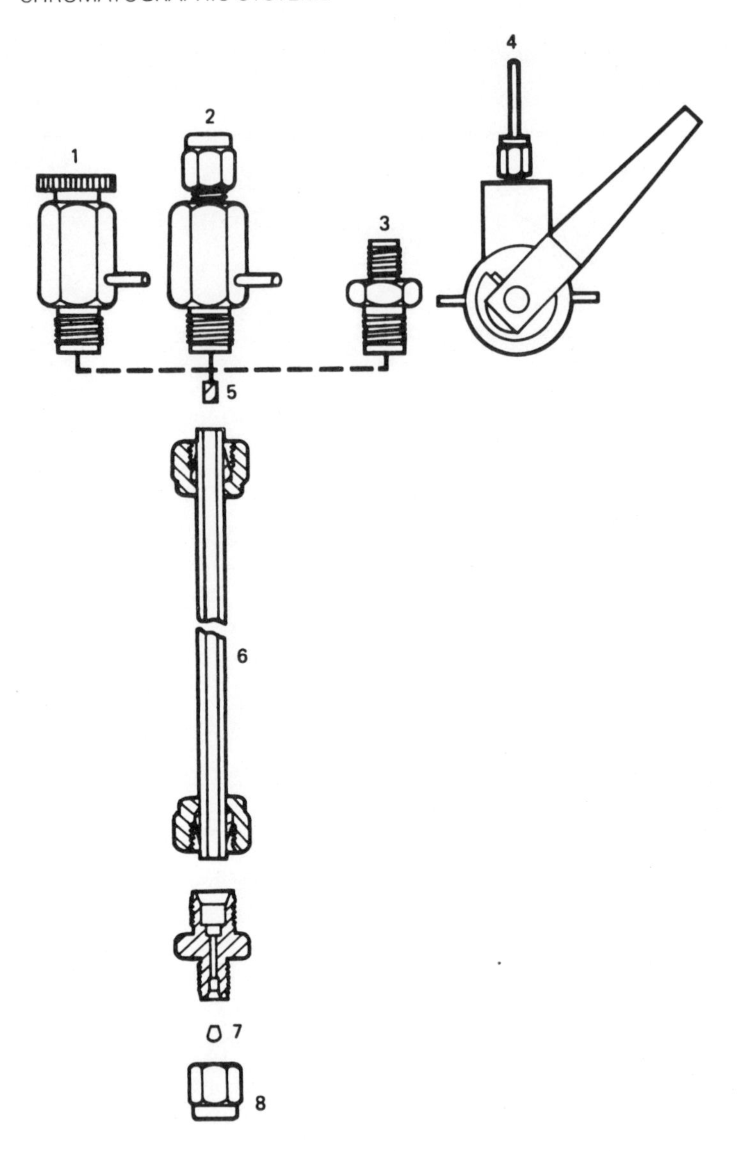

Fig. 3-1. Sample injector designs for liquid chromatography. (1) On-column septum injector; (2) stop-flow injector; (3) reducing union with bonded frit; (4) loop injector; (5) Teflon filter retainer; (6) precision-bore, heavy wall column, 316SS; (7) silver-plated ferrule; (8) 3166SS nut. (Courtesy of Perkin-Elmer Corporation)

The more commonly used septa and recommended solvent applications are seen in Table 3-1. This table is presented because the improper combination of septum and solvent can produce undesirable results. Major problems encountered are those of septum/solvent interaction by "leaching" materials (generally plasticizer) from the septum and hardening of the septum, limiting its useful application to one or two injections before it begins leaking. The "leaching" phenomenon will become quite evident and is a perfect analogy to gas chromatography septum "bleed" as observed on the recorder trace. A gradual increase or "drifting baseline" will be the result. Because the detector has a low dead-volume this added signal will result in a decrease in the sensitivity of the components being sought in the analysis. The linear range of the detector for the components of interest will also be affected unless accommodations are made for the lower sensitivity. Subsequently, septum/solvent interaction results in hardening of the septum. In many cases, the septum actually becomes brittle and even without syringe injection may begin leaking, especially at high pressure. Disruption of the flow characteristics within the system is inherent with the changing of the septum and, generally, re-equilibration of the system takes longer for LC than for gas chromatography. In changing the septum, precautions should be taken to insure that only minimal air is introduced into the system as this will merely prolong equilibration time. This is best done by reducing the flow of solvent delivered by the pump. Remove septum and adapter, change septum and slowly re-mount the adapter, allowing excess solvent to flow into the adapter and displace any air which may become entrapped. Then tighten the adapter and resume normal operation.

The technique of syringe injection also merits a few comments. A fixed-needle type syringe is recommended. The needle should be tapered and free of barbs, otherwise the useful life of the septum is decreased. After filling the syringe and before injecting the sample, any air entrapped in the syringe may be removed by inserting the needle of the "loaded" syringe into a piece of silicone gum rubber and slowly depressing the plunger to expel entrapped air. High pressure syringes may be purchased from Hamilton Co., Precision Sampling Co., Unimetrics, SRI, and Glenco Scientific Co.

Stopped-flow or interrupted flow injection has been used as a sample introduction technique. It is best used, however, with syringe pumps. Designs are available which prevent virtually any disruption in the system and allow minimal introduction of air, providing the preceding precautions on syringe filling have been employed. Syringe

TABLE 3-1

Recommended choice of septa for use with various solvents

Solvent	Recommended Septum
Methylethyl Ketone	EPR
Tetrahydrofuran	EPR
Dimethylformamide	BUNA-N
Alcohols	BUNA-N
Toluene	Viton-A
Benzene	Viton-A
Trichlorobenzene	Viton-A
Cresols	Viton-A
n-Hexane	BUNA-N or Viton-A
Water	EPR, BUNA-N, or Viton-A
Most Solvents	White Silicone Gum Rubber

injection is used for sample introduction. For quantitative applications, the stopped-flow technique is best utilized by incorporating an internal standard with the mixture to be analyzed. Necessary design features are minimal dead-volume and cleanly swept flow geometry.

Loop injectors, which are used in almost all high molecular weight GPC (Gel Permeation Chromatography) applications, are becoming widely used in high resolution liquid chromatography also. Two features of these injectors had to be improved for high resolution work. The loop injectors used for normal GPC work had too much dead-volume for high resolution LC use. Secondly, this type injector must operate over a wider range of pressures. Currently, thereare several suppliers of loop injectors which are compatible with high resolution LC demands. One such injector is shown in Figure 3-2 (2). This design permits sample loading without septum injection and without stopping the flow of eluant. Reproducibility is good because the sample is introduced into the system at atmospheric pressure using a syringe. The injector will accommodate virtually any size sample (1 μl to 2 ml. or larger) rendering it applicable to both analytical and preparative LC. Operating pressure capability to 6000 psi satisfies high-speed LC demands.

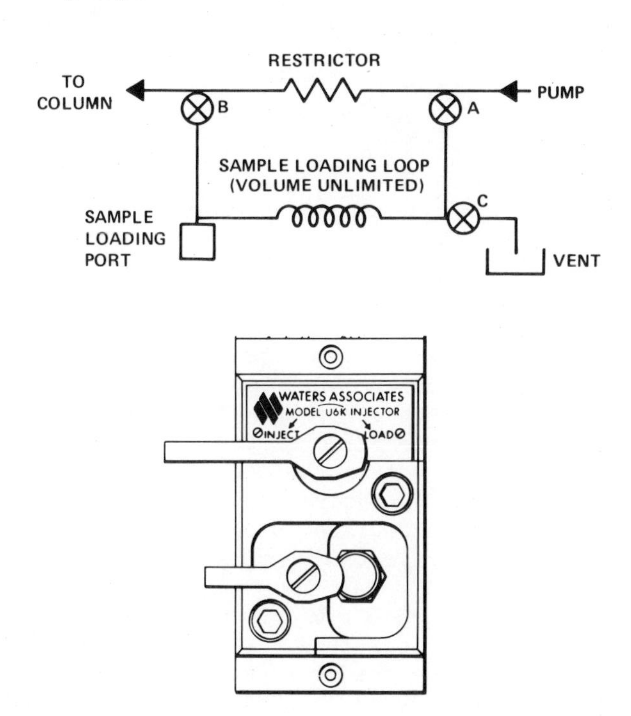

Fig. 3-2. Universal loop injector (U6K, Courtesy of Waters Associates Inc.) Start: *Valves A and B are closed in the "Load" position allowing flow through the restrictor to column.* Sample loading: *Sample is introduced into sample loading port and lever switch is turned to the "Inject" position opening valves A and B. The restrictor causes flow to pass through the loop transporting sample to the column.*

Another design of loop injector is seen in Figure 3-3 (3). This valve is also quite durable and satisfies the requirements described earlier. "Load" and "Inject" positions again permit sample introduction to be made at atmospheric pressure.

In any case, the principle involved in loop injectors is that of syringe-filling a small sample loop with sample and by valve-switching the sample in the loop is displaced by diverting the flow of mobile phase through the loop and onto the column. Ideally, the sample to be analyzed should be dissolved in portions of the mobile phase to eliminate unnecessary solvent peaks.

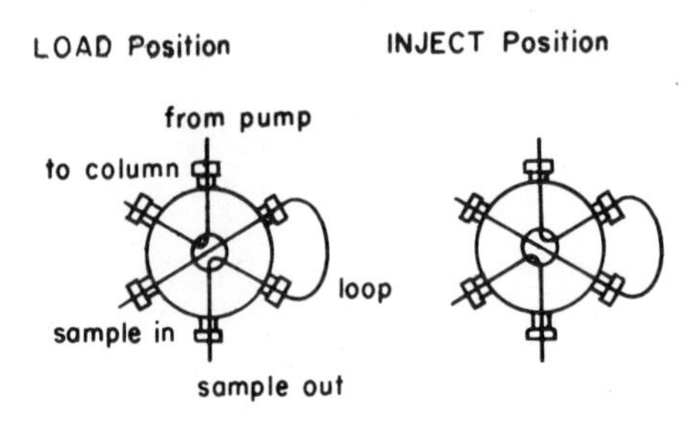

Fig. 3-3. Sample injection valve. Samples are injected with this valve in two stages. With the valve in the "Load" position, as diagrammed at left, the sample is drawn with a syringe from a vial into the sample loop, which is off-stream at this stage. Then the valve is turned to the "Inject" position, thereby connecting the loop into the stream from the pump to the column. Because only the liquid in the loop enters the stream, consecutive injections are identical in volume. (Courtesy of Spectra Physics Corp.)

For high-resolution LC, the loop injector seems superior to the other types of injectors. Design of current loop injectors has created low dead-volume, high-pressure capability. These features are required when narrow-bore columns containing microparticulate packings are used.

Another type injector is shown in Figure 3-4 (4). This is a pneumatically-operated sample injector and is "loaded" at atmospheric pressure using a suitable syringe. Sample size is controlled by a calibrated volume groove in the sliding pneumatic shaft. To vary sample size, one changes to a shaft with a different groove size. These valves are designed only for analytical work. In reassembly, shaft alignment is critical to proper function and good reproducibility. Referring to Figure 3-4, if the valve is mounted in a horizontal configuration, the drain (D) should be extended with a length of tubing to form a "U"-shape. The length of the tube should be such that the end extends above the level of the sliding pneumatic shaft. This becomes particularly important when samples to be analyzed are dissolved in liquids of low viscosity and low-surface tension. Otherwise, sample introduced into the groove will leak due to gravitational flow and result in non-reproducibility of the sample quantity injected.

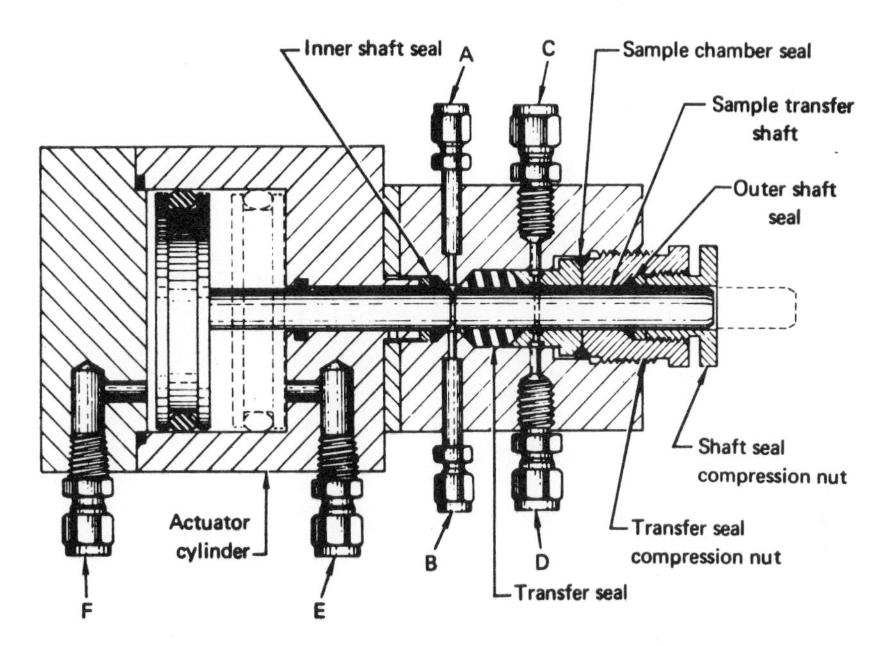

Fig. 3-4. High pressure sample valve for liquid chromatography usable at pressures to 5000 psig. Connections are: (A) Mobile phase, (B) Chromatographic column, (C) Sample, (D) Drain, (E) and (F) Solenoid actuated 2-way pneumatic valve. Air source is 50-100 psig. (Courtesy of Hamilton Company)

The sample injection system must be kept clean. Again, this is best accomplished via miscible solvents and may be carried out simultaneously with pump cleaning. The use of corrosive solvents should be eliminated as most injectors will be damaged by such treatment. Buffered solutions should be flushed from the system before leaving in a standby condition overnight.

Sometimes particulate matter from solvents and pump seals, etc., may become lodged in the small diameter tubing leading to the injector or in the small valve orifice associated with the injector inlet line from the pumping system. This produces several problems. Pumps having a pressure read-out device will show an increase in pressure due to this restriction in the transfer line to the injector. Analysis under these conditions also produces undesirable results. The components separated on the chromatographic trace will lead the observer to believe that perhaps the column efficiency has decreased in some fashion. Generally, these symptoms are peak broadening and tailing peaks. When this occurs one should not be too eager to suspect or discard the column without carrying out a few check procedures. 1) Observe the system pressure (switch injector from "Load" to "Inject," observe ΔP). 2) Measure the flow rate of mobile phase at the detector exit. 3) Inspect all fittings and valves for leaks in the system. Usually, if the column loses efficiency, it will be a more gradual occurrence than that observed from a partially plugged injector or transfer line. If, in fact, the injector or transfer line to the injector is partially plugged, a good remedial action is that of reversing the flow of mobile phase through these components. Merely attach the pump effluent line to the injector *outlet* and backflush the injector and associated transfer lines from the pumping system. After five or ten minutes of backflushing the system at a reasonably high flow rate, i.e., 3-5 milliliters per minute, connect the system in the normal flow scheme and observe the system pressure and flow rate. If the system pressure reading is still higher than normal suspect either a plugged column frit or an inaccurate reading from the pressure transducer in the pumping system.

Occasionally, septum injectors will become clogged after repeated injections. Generally, this plugging is due to small particles of the septum being deposited in the injector. Reversal of flow through the injector will usually remove the plug.

The fittings and transfer tubing from the injector exit to the column inlet is also critical. These connections should always be made with the thought of introducing only minimal dead-volume in this vicinity. If dead-volume is excessive the sample introduced will not be deposited in a narrow band onto the column. Band broadening is the result; and, in the case of low concentration components in the mixture, a subsequent loss in sensitivity is obtained from the passage of broad diffuse peaks through the detector.

In conclusion, the loop injector or valve seems preferable for high resolution LC applications because current designs achieve low dead-volume and less disruption in flow equilibrium.

REFERENCES

1. Septum Injector Designs, Perkin Elmer Corp.

2. Universal Injector (U6K), Waters Associates Inc.

3. Sample Valve for Liquid Chromatography, Spectra Physics Corp.

4. High Pressure Pneumatic Sample Valve, Hamilton Corp.

QUESTIONS AND ANSWERS

Question: What should one look for in choosing an injector or valve for HSLC?

Answer: It is generally felt that a sample valve provides much better reproducibility provided it is cleanly swept and does not impose but minimal flow disruption through the column and detector during sampling. In either case, syringe injector or valve, it should be of minimal dead-volume and constructed such that it will withstand the pressure requirements of the problem. Another point which should be considered before purchasing separately a sample introduction system is to be sure that the associated fittings are compatible with your column fittings.

Question: Approximately how many injections can one expect to make in HSLC before it becomes necessary to change septa?

Answer: Assuming the syringe is free of barbs and the proper septum/solvent choice has been made, as many as ten to twenty injections may be carried out. This is also dependent upon the pressure imposed upon the septum.

Question: If one wishes to increase sensitivity or introduce larger quantities of sample onto the column, how is this best accomplished?

Answer: It is preferable to inject larger quantities of a dilute solution rather than smaller quantities of a concentrated solution. In so doing, the column capacity is increased and band broadening is less with larger volumes of dilute solutions.

Chapter 4

LC COLUMNS (SELECTION AND PREPARATION), COLUMN OVENS, AND COLUMN HEATERS

Selecting the proper mode or method of liquid chromatographic separation is not always straightforward. Table 4-1 provides a convenient guide for establishing the type separation required for most systems encountered in liquid chromatography. A more fundamental description of the various type (liquid-liquid, liquid-solid, ion exchange, and gel permeation) is found in Chapter 1, "Introduction to Liquid Chromatography."

Stainless steel or glass columns are commonly used. For pressures exceeding 500 psi, metal columns must be used. Currently available glass columns and associated fittings have a quoted pressure limitation of 500 to 1000 psi. Karger and Barth (1) have concluded that there were only small differences in column efficiencies for stainless steel, aluminum, copper or glass columns; whereas, teflon-coated aluminum tubes were found to yield efficiencies that were poor and non-reproducible. In their study, they also concluded that wall effects (active sites and roughness) were insignificant for the systems evaluated.

The diameter and length of the column chosen is governed by the mode and efficiency of separation required. For gel permeation chromatography, standard-wall 3/8 inch o.d. or 1/4 inch o.d. stainless steel or glass columns are used. For high resolution liquid-liquid, liquid-solid, or ion exchange applications where the pressure drop along the column

Note A

Aqueous Systems

Gel Filtration (GFC)

1. Choose smallest mesh possible to attain high efficiency.
2. Choose correct porosity according to molecular weight range of interest.
3. Avoid excessively high pressure.

Note B

Non-Aqueous Systems

Gel Permeation (GPC)

1. Column materials recommended:
   a) cross-linked polystyrene
   b) polyvinylacetate
   c) porous glass
2. Solvents:
   a) If UV detection is used, choose a UV transparent solvent.
   b) For maximum sensitivity when using RI detectors, choose solvents which reflect the maximum RI compared to the species being separated.

Note C

Ionic Species

Ion Exchange (IEC)

1. Decide which type of resin to use, i.e., anion or cation.
2. Always use resin of smallest particle size to obtain maximum efficiency. For analytical scale work, use pellicular resins.
3. Other than column choice the most important variables are pH and ionic strength of the eluant. For anions $pH>7$, for cations $pH<7$. Ionic strength is increased to elute more strongly bound species.
4. Gradient elution is quite useful for optimizing the system.

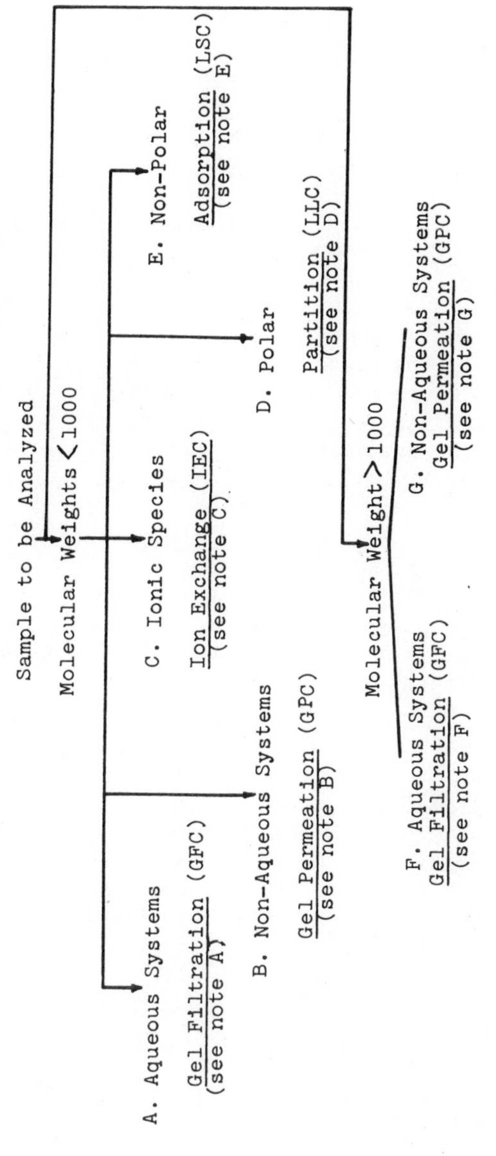
Table 4-1
CHOOSING THE CORRECT MODE OF LIQUID CHROMATOGRAPHY
Sample to be Analyzed
Molecular Weights < 1000
A. Aqueous Systems
Gel Filtration (GFC)
(see note A)
B. Non-Aqueous Systems
Gel Permeation (GPC)
(see note B)
C. Ionic Species
Ion Exchange (IEC)
(see note C)
D. Polar
Partition (LLC)
(see note D)
E. Non-Polar
Adsorption (LSC)
(see note E)
Molecular Weight > 1000
F. Aqueous Systems
Gel Filtration (GFC)
(see note F)
G. Non-Aqueous Systems
Gel Permeation (GPC)
(see note G)

Note D

Polar

Partition (LLC)

1. A non solvent must be chosen for the liquid phase or the solvent must be saturated with liquid phase.

2. If a bonded liquid phase is used, solvent programming may be employed.

3. UV detectors may be used to eliminate drift if a low optical density solvent is used.

Note E

Non-Polar

Adsorption (LSC)

1. Use TLC solvent screening.

2. If more than one solvent is needed to obtain the separation and RI detection is used, choose solvents with approximately the same RI.

3. If the species are UV absorbers and a UV detector is being used, choose solvents which are transparent in the UV region. Be aware of UV absorbing inhibitors which are added to non-UV absorbing solvents.

Note F

Aqueous Systems

Gel Filtration (GFC)

1. Use low porosity gels. (See Aqueous Systems, Molecular Weights > 1000)

Note G

Non-Aqueous Systems

Gel Permeation (GPC)

1. Use low porosity gels. (See Non-Aqueous Systems, Molecular Weights > 1000)

is very high, 2 to 4 millimeter i.d., 1/4 inch o.d. and 1/8 inch o.d. or 1/16 inch o.d. (thin-wall) stainless steel tubing is used. Column lengths for GPC applications are necessary longer than those needed for other modes of liquid chromatography. Effective column lengths for GPC applications equivalent to twenty feet are not uncommon especially if the solvent transport system (see Questions and Answers Chapter 2) has recycle capability. LLC, LSC and ion-exchange chromatographic column lengths may range from less than one foot to 10 feet in length.

The shape or geometry of the packed column has received much attention in correlation with column efficiency (2). Various configurations have been used; however, straight columns are preferred because the efficiency obtained during packing is preserved in use and better reproducibility can be obtained among columns which are packed straight and used as such.

Many column compartments will accommodate only certain lengths of columns. Several straight columns may be used in series in such cases by "coupling" the column ends with short (one to three inches) lengths of small diameter, low dead-volume tubing. One-sixteenth inch (o.d.) and three to four one-hundredths inch (i.d.) stainless steel tubing is used. The fittings associated with these couplings should always be of low dead-volume design. In some cases, portions of the packing material are placed into the dead-space of the union eliminating the voic area. The tube coupling is semi-circular in shape.

Pre-packing with subsequent coiling and/or bending of longer columns has provided satisfactory results. Configurations of this type are used with one-eighth inch (o.d.) narrow-bore (1-2 mm i.d.) or one-sixteenth inch (o.d.) thin-walled capillaries. Some workers (3,4) claim that round and sharp bends in LC columns have advantages by decreasing the resistance to mass transfer of sample components passing through the column. Presumably, small voids and packing inhomogeneities are formed within the column which are virtually impossible to reproduce. If these columns are not excessively long, the overall efficiencies may be poor, yet reproducible, within certain limits. There are no advantages realized in reproducing poor efficiencies. For example, in using a narrow-bore column ten feet in length having two or three sharp bends and correspondingly two or three voids, it follows that column efficiency may be drastically impaired. In spite of the advantages obtained via decreasing the resistance to mass transfer, voids contribute to eddy currents in the column with subsequent recombinations of separated components and band broadening (5) which, for high resolution liquid chromatography, is intolerable. One could undoubtedly achieve the same or better efficiency using a straight column three feet in length.

Methods of packing columns for liquid chromatography vary widely depending upon the nature of the material to be packed. Most columns for liquid-liquid or liquid-solid chromatography are merely dry-packed (37 to 70 μm particles) using a tamping or "bumping" technique (6), (21). Porous ion exchange resins must be wet-packed (7) in the form of a slurry, allowing adequate time for swelling and particle distribution between each addition of the packing. It is recommended that once these columns are packed in a wet slurry they should be kept this way when not in use by capping the ends of the column.

GPC columns are also pressure-packed in the form of a slurry (8). Again, as with conventional ion-exchange resins, once packed they should not be permitted to go dry. If the columns lose liquid and become dry, contraction of the gel occurs creating voids in the rather large diameter column. Subsequently, the column is virtually ruined because even replenishing the liquid lost or the addition of more packing material does not restore the column to its initial efficiency.

GPC columns are generally purchased pre-packed from the manufacturer. These columns are quite expensive and rightfully so because few workers in the field can reproduce GPC columns packed in their own laboratories. For this reason alone, it is recommended that GPC columns be purchased prepacked. This has its advantages in spite of the costs involved. Generally, the supplier will provide information regarding the efficiency or plate-count for any given set of columns purchased. The theoretical plate count varies in accordance with the packing material, column diameter, and performance tolerance of the column. When ordering GPC columns, provide separation only for the molecules you wish to separate. Do not specify columns with larger exclusion limits than the limit required by the larger molecules you wish to separate. Most GPC columns ordered have a specified capability of molecular weight applications. Generally, this capability is expressed in Angstrom units, a rather arbitrary designation. Even so, this designation has useful application within a factor of two, regarding molecular weight. With this in mind, then, the GPC Angstrom size of a molecule is the molecular weight divided by 20. Generally, molecules which have a high density per unit molecular length, such as polystyrene, have a higher multiplier, and molecules which have a fairly low density per unit chain length, such as polyethylene or polypropylene, have a multiplier as low as 15. The GPC Angstrom size is defined as 1/41 of the molecular weight of polystyrene (9). Some things to keep in mind when evaluating GPC columns for efficiency and reproducibility are:

a) a change in the volume of the system between the injector and the column.
b) a change in the volume of the system between the outlet of the column and the detector.
c) a change caused by the density of packing in the column (different solvents, etc.).
d) a change caused by the density of the pores and pore-size distribution within the polymer head, i.e., different solvents, pressure, etc.

Each of these changes, individually, is small; however, a combination of any of these can result in considerable deviation from any standard curve. There is information available on current state of the art in GPC (10), as well as methods for optimizing resolution (11) in this important area. It behooves one, then, to evaluate his particular LC system and be aware of possible anomalies which may occur.

In contrast to the earlier thinking of some workers, column heating and column ovens may drastically influence the efficiencies ofsome applications in liquid chromatography. For GPC applications the effects are minimal; however, for high-speed liquid chromatography the results can be significant. This is especially true for high-speed ion exchange liquid chromatography whereby the ion exchange process may be markedly enhanced by additional heat applied to the column. In the liquid-liquid chromatography mode, column heating is especially significant. All applications in this area rely upon the compatible solubility of components to be separated in the liquid or stationary phase. This solubility is inherently temperature dependent. Virtually all of the liquid-coated materials have maximum temperature limitations and these limitations should be adhered to in practice. The use of a pre-column or an eluant containing a portion of the coated liquid phase is more essential than at ambient temperature for replenishing the loss of liquid phase when a sufficiently high temperature is used on the column. The detection system used should be compatible with the additional heat applied. As in gas chromatography, cold spots in the column or cold spots between the column exit and the detector may produce undesirable and non-reproducible results. The ideal approach, applicable to all detection systems, is that of a thermostated column oven, whereby the column and detector may be maintained at approximately the same temperature. If this is not possible a heat exchanger should be incorporated in the detector housing. The output of a differential refractometer is obviously temperature dependent, whereas an ultra-violet detector yields only minimal response to small changes in temperature. Column jackets, which control column temperature accurately via circulation of water through a jacket, are very efficient and are used quite extensively with UV detectors.

Table 4-2 (20) shows the effect of temperature variation with constant flow rate on peak areas. This is based upon data obtained from the analysis of a simple fused-ring aromatic mixture (Chapter 2, Figure 2-9) whereby the percent standard deviation of absolute areas is used. The system was studied over a 4°C temperature range at 58°C and 29°C.

TABLE 4-2

Effect of temperature (t) and flow rate (v) on peak areas constant flow rate, variable temperature (20)

Absolute Areas	
σ, 58° → 54°C	σ, 29° → 25°C
5.0%	4.1%
Normalized Areas	
σ, 58° → 54°C	σ, 20° → 25°C
0.3%	0.2%

The data substantiates the need for temperature control. In this study, only the column was heated and UV detector was at ambient temperature. Further experiments have shown that incorporating a heat exchanger (small inside diameter stainless steel tubing) in the flow scheme prior to introducing the column effluent to the flow cell reduces the percent standard deviation. This indicates that operating the column and detector at the same temperature results in only minimal deviation in absolute areas among successive analyses.

There have been many improvements in column packing materials which yield higher efficiencies (12). Excellent review articles (13, 14, 15) describe the development and use of these materials for HSLC. Much emphasis has been placed upon the small diameter silicas and aluminas. These are porous particles ranging in size from 5 to 15 micrometers (μm) as the particle diameter (dp). Particles are either irregular shaped (15) or spherical (16). Due to the porous nature of these materials, sample capacity is large. Efficiencies of ten to twenty thousand theoretical plates per meter are not uncommon. Preparation and packing of these small dp's are accomplished by slurrying with the appropriate solvents (15) and subjecting the slurry to high pressure packing, i.e., four to six thousand psi.

Preparative applications in HSLC are commonly feasible. The use of short but large diameter columns may provide efficiencies comparable to analytical mode operation (17). The parameters considered are those of column efficiency and resolution with variations in sample volume, sample concentration, solute weight and mobile phase linear velocity.

Snyder (18, 19) has proposed a two-part series for estimating sample resolution and column length which is quite useful in designing or optimizing a separation scheme. Some information is provided regarding quantitation and resolution of adjacent peaks. Other variables which must be considered are reproducibility of the pumping system, sample injection techniques, and detector sensitivity and linearity (20).

In conclusion, liquid chromatography separations will improve almost proportionally to the user's knowledge and experience with his particular unit with regard to choice of separating modes, proper column choices and the use of controlled and equilibrated column heating.

REFERENCES

1. B. L. Karger and H. Barth, *Analytical Letters*, Vol. 4, No. 9, September 1971, p. 602.

2. H. Barth, E. Dallmeier, and B. L. Karger, *Anal. Chem, 44*, p. 1726, September 1972.

3. R. P. W. Scott, D. W. Blackburn, and T. Willing, *J. Gas Chromatography*, *5*, 183, 1967.

4. J. J. Kirkland, *J. of Chromatographic Science*, *7*, 361, 1969.

5. E. Grushka, *Anal. Chem.* (46), p. 510A, May 1974.

6. D. Randau and W. Schnell, *Journal of Chromatog.*

7. C. D. Scott and N. E. Lee, *J. of Chromatography*, *42*, 263, 1969.

8. K. J. Bombaugh in *Modern Practice of Liquid Chromatography*, J. J. Kirkland, Ed., John Wiley and Sons, Inc., New York, 1971.

9. *Chromatography*, Waters Associates, September 1970.

10. A. R. Cooper, J. F. Johnson and R. S. Porter, *American Lab.*, May 1973, p. 12.

11. J. F. Johnson, A. R. Cooper and R. S. Porter, *Journal of Chrom. Science*, June 1973, p. 292.

12. F. R. MacDonald, *American Lab.*, May 1973, p. 80.

13. R. E. Leitch and J. J. DeStefano, *Journal of Chrom. Science*, Vol. 11, March 1973, p. 105.

14. R. E. Majors, *American Lab.*, pp. 27-39, May 1972.

15. R. E. Majors, *Anal. Chem.* (44) 1722, 1972.

16. Phase Separations Ltd. Bulletin, Spherisorb, 1974.

17. J. J. DeStefano and Beachell, *Journal of Chrom. Science*, p. 654, Nov. 1972.

18. L. R. Snyder, *Journal of Chrom. Science*, Vol. 10, pp. 200-212, April 1972.

19. L. R. Snyder, *Journal of Chrom. Science*, Vol. 10, pp. 369-379, June 1972.

20. R. A. Henry, M. T. Jackson, S. Bakalyar, "Quantitation in Liquid Chromatography" presented at 1974 Pittsburg Conference on Anal. Chem. and Applied Spectroscopy.

21. I. Halasz and M. Naefe, *Anal. Chem.*, Vol. 44, p. 76, 1972.

QUESTIONS AND ANSWERS

Question: Why does dead volume in the injector and column ends have more detrimental effects in LC than GC?

Answer: Because in LC, the diffusion rates for liquids is 10^4 - 10^5 times less than for gases and, hence, the time duration of solutes in the void or dead volume area causes band broadening which lowers the efficiency.

Question: Of the modes possible in high speed LC, namely adsorption, partition, reversed-phase and ion-exchange, which one seems to yield better quantitation?

Answer: This depends largely upon the analyst. It is felt that reversed-phase applications yield better data for quantitative purposes because adsorption effects and "ghosting" on the stationary phase surface is minimal compared to adsorption. Both adsorption and partition LC can be equally as good if (a) in the case of adsorption, the silica or alumina surface is adequately and reproducibly deactivated and (b) in the case of partition, the amount of coated or "bonded" stationary phase remains constant. Ion exchange can provide precise and accurate data provided the resin is active and good buffer control (pH and ionic strength) is adhered to. All things considered, regardless of the LC mode, one should incorporate the use of an internal standard when the maximum in quantitative output is desired.

Question: What are considered to be "major break-throughs" in modern-day LC column technology?

Answer: Speed and efficiency obtained via the use of pellicular or superficially porous particles and, the development of sizing and packing techniques associated with extremely small particles in the range 5 - 10 micrometers. Also of equal importance is that of permanently bonding stationary phases to supports for producing durable and stable column packings.

Chapter 5

LC DETECTION SYSTEMS

The general purpose of detection systems in liquid chromatography is evident. The choice of a suitable detector, many times, is not quite as evident. The liquid chromatographer has several means of detecting what he has so laboriously separated. Therefore, the choice of detection is important to the overall scheme of the LC make-up and is contingent upon such criteria as the noise, sensitivity and linearity requirements of the problem.

The two basic categories of detectors are solute property detectors and bulk property detectors. The solute property detector is sensitive to some physical property of the component being analyzed and is relatively insensitive to the eluant or mobile phase. Examples of this type are UV absorption, fluorimeters, radioactive detectors, solute transport detectors, and polarographic detectors. Bulk property detectors monitor changes in the physical properties of the mobile phase. Examples are the differential refractometer, conductivity and dielectric constant detectors. The general principles of some of these detectors are described; however, special emphasis is placed on the more common modes of detection, i.e., UV and RI.

The conventional ultra-violet detector records the absorbance of a component or molecule at a fixed, specific wavelength. The more commonly fixed wave-length detectors available are 254 and 280 nanometers (nm). This detector offers great sensitivity, i.e., 0.0001 absorbance units and a linear dynamic range of approximately 5 x 10^3.

The source of energy for these UV detectors consists of a low pressure, hot cathode mercury lamp. Because of the penetrating power of UV radiation, a word of caution . . . . with the naked eye, do not look directly into the cell or source lamp while in operation! One should always wear safety glasses or other eyeglasses.

The UV detector is not universal and must be accepted as such because many materials do not exhibit UV absorption. This limits its utility to compounds which absorb UV radiation. Byrne (1) stated that non-UV absorbing compounds may produce signals in the UV detector as a result of changes in refractive index due to light scattering phenomena.

There are numerous suppliers of UV detectors and all function on basically the same principle with some minor differences. The principle of operation of the UV detector consists of light sensitive resistors arranged in a Wheatstone bridge circuit and the sensitivity of these resistors is universely proportional to the UV light impinging upon them. When a UV absorbing component enters the cell from the column an imbalance is recorded as a peak on the chromatogram.

Figure 5-1 shows a diagram of a typical UV detector.

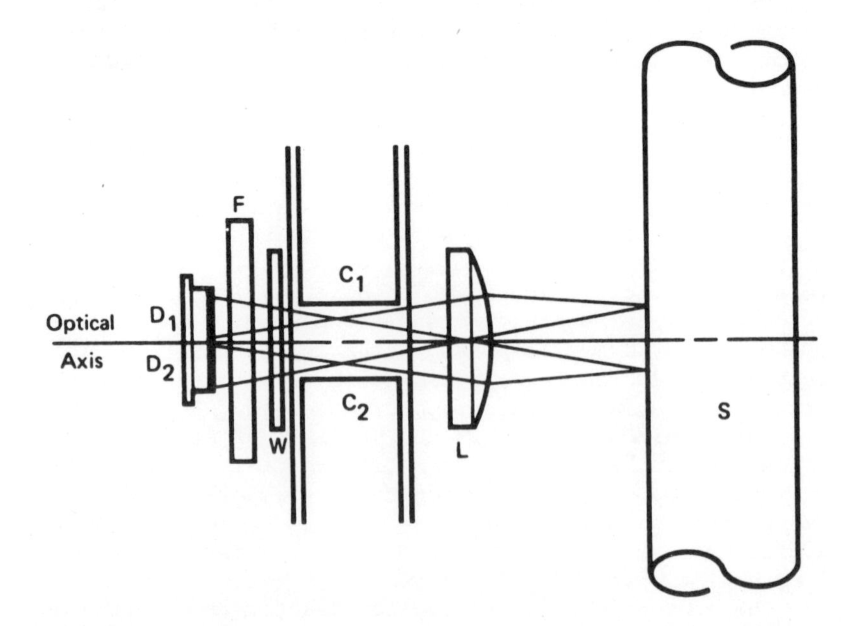

Fig. 5-1. Optical path of a typical UV detector. (Courtesy of Laboratory Data Control, a Div. of Milton Roy Co.)

Two beams of radiation from a common area on the source lamp, S, pass through lens, L, flow cell chamber, C1' and C2, plane window, W, a visible light blocking filter, F, and finally impinge upon dual photodetector sensitive areas, D1 and D2. The radiation passing through the chambers is collimated. As the cell chamber axes are mutually parallel, the result is that the two beams passing through the chambers have originated from virtually the same area on the source lamp, S. The use of a common source area for the two beams is advantageous because spatial variations in lamp brightness as a result of temperature changes, dust accumulation, droplets of mercury condensate, etc., are "common mode" to both beams and do not affect the detector readout.

In addition to its low cell volume (10μl), which does not degrade chromatographic separations, the system is inherently stable due to a single optical axis and a single set of optics common to both beams.

Some UV detectors have a meter or digital output displaying the bridge excitation voltage. With the bridge voltage balanced, this meter should read within certain limits as described in the manual for that detector. The function of this meter is to assure the user that sufficient radiation is reaching the photodetector elements. The output of this meter will increase in the presence of large signals. This meter is an extremely valuable troubleshooting device. Air bubbles passing through the detector may be diagnosed quite readily by observing the meter output.

An air bubble build-up may also be observed. A meter output higher than normal indicates that the eluant (solvent) may contain a small amount of an absorbing material (s), perhaps added to the solvent by the supplier to function as an inhibitor. A classic example is that of the solvent tetrahydrofuran (THF) containing small amounts of the inhibitor ionol (BHT).

The bridge excitation meter may also signal the user that the detector needs cleaning. Again, higher than normal output on the meter may indicate component film deposits within the cell, which may have occurred during previous runs with different solvents and solutes. The recommended approach is that of cleaning with a miscible solvent series or replacing the cell windows.

A choice of two detection wavelengths is commonly available for most UV detectors. The more common choice is 254 nm; however, there are numerous compounds which exhibit greater sensitivity at 280 nm or other wavelengths. This choice is provided by UV filters which are easily changed. In humid or dusty areas these filters may become coated either with condensate or dust resulting in erratic recorder response or a decrease in sensitivity. A periodic check of these filters subsequently wiping with a non-abrasive, lint-free cloth or napkin, will eliminate these problems

Dual wavelength UV capability has been commercially available for at least five years (2, 3). The common wavelengths monitored are 254 and 280 nm. This was accomplished either by tandem optic units and flow cells specific for each wavelength (2) or by a single optic unit with dual or mixed wavelength capability (3). A schematic diagram of the latter is seen in Figure 5-2.

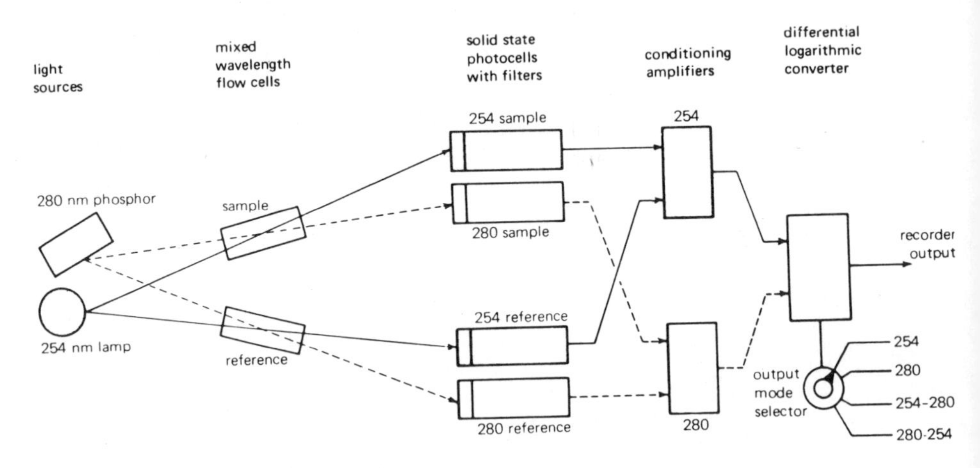

Fig. 5-2. Schematic diagram of a mixed wavelength UV detector. (Courtesy of Spectra Physics)

The 254 (λ1) low pressure mercury lamp radiation shines through both sample and reference flow cells onto sample and reference photocells. The 280 nm (λ2) phosphor (cerium-activated lanthanum fluoride) shines through the cells simultaneously. The cells act as apertures which separate the 254 nm and 280 nm radiation, directing them to the different photocells. Appropriate filters are placed over the photocells. This optical configuration can yield four output modes, i.e., λ1, λ2, λ1 - λ2, and λ2 - λ1. Preventive maintenance includes an occasional check of the λ1 and λ2 balancing voltages within the optic unit and inspection of the cell windows for film deposits.

A classical problem in absorbance detection is that of distinguishing between true sample (component) output versus anomalies which occur due to refractive index phenomena of pure solvents within the flow cell. Reflection of UV light from the cell wall is negligible; however, deflected light which strikes the cell is absorbed and appears as an absorbance change on the detector even in the absence of any sample component. If one considers a typical UV detector flow cell there are two types of refractive index changes which may occur. 1) From the source lamp, light is emitted which, upon entering the cell and striking the liquid interface, is bent because of the refractive index difference between the quartz cell window and the solvent. 2) Within the cell there are transients occurring due to changes in flow rate, solvent gradients and sample components.

In any case, and specifically in gradient elution where two liquids are being continually mixed in varying proportions (depending upon the gradient profile chosen), appreciable changes in refractive index do occur. Considering that, generally, when two liquids are mixed heat is evolved and the refractive index property is a function of temperature, we may conclude that anomalous peaks will appear even in a "blank" gradient run. These peaks will not be small in magnitude but will be governed, to a greater degree, on the rate of mixing and the concentration limits and time of the gradient analysis.

A new development in detector cell geometry has greatly reduced these effects of anomalous peaks during gradient elution analysis. The "tapered" cell geometry shown in Figure 5-3 (4) diagrams the analogy. This design insures that all of the emitted light which enters the cell will leave the cell if there is no true absorbance; therefore, anomalous peaks caused by mere solvent mixing are greatly reduced at no loss in sensitivity. This detector offers not only dual wavelength monitoring, but two separate flow cells for dual system monitoring (Figure 5-4 (4), if desired.

One inherent disadvantage of all the commercially available dual wavelength detectors is that there is no monitoring capability below 254 nm. Much interest and application recently requires monitoring below 254 nm, i.e., 200-210 nm where detectors employing a deuterium lamp source and variable wavelength capability have satisfied this need.

Recently, multiple or variable wavelength detectors (5, 6) have become available for high-speed LC applications. These detectors greatly increase the scope of compounds which may be analyzed and also may enhance the sensitivity compared to fixed wavelength detectors. At least one design (5) permits stopping a chromatographic peak or component in the flow cell and subsequently scanning the range of interest. This necessitates turning off the chromatographic pump. The time required for re-equilibration of the flow scheme is dependent upon the type pump used to a greater extent.

These detectors are capable of studying compounds in the range 210 to 630 (6) or 780 mm (5). As thermal turbulence is a major cause of noise, the stability of lamp sources, power suppliers and signal electronics become extremely critical. Generally, sensitivities of components obtained using single

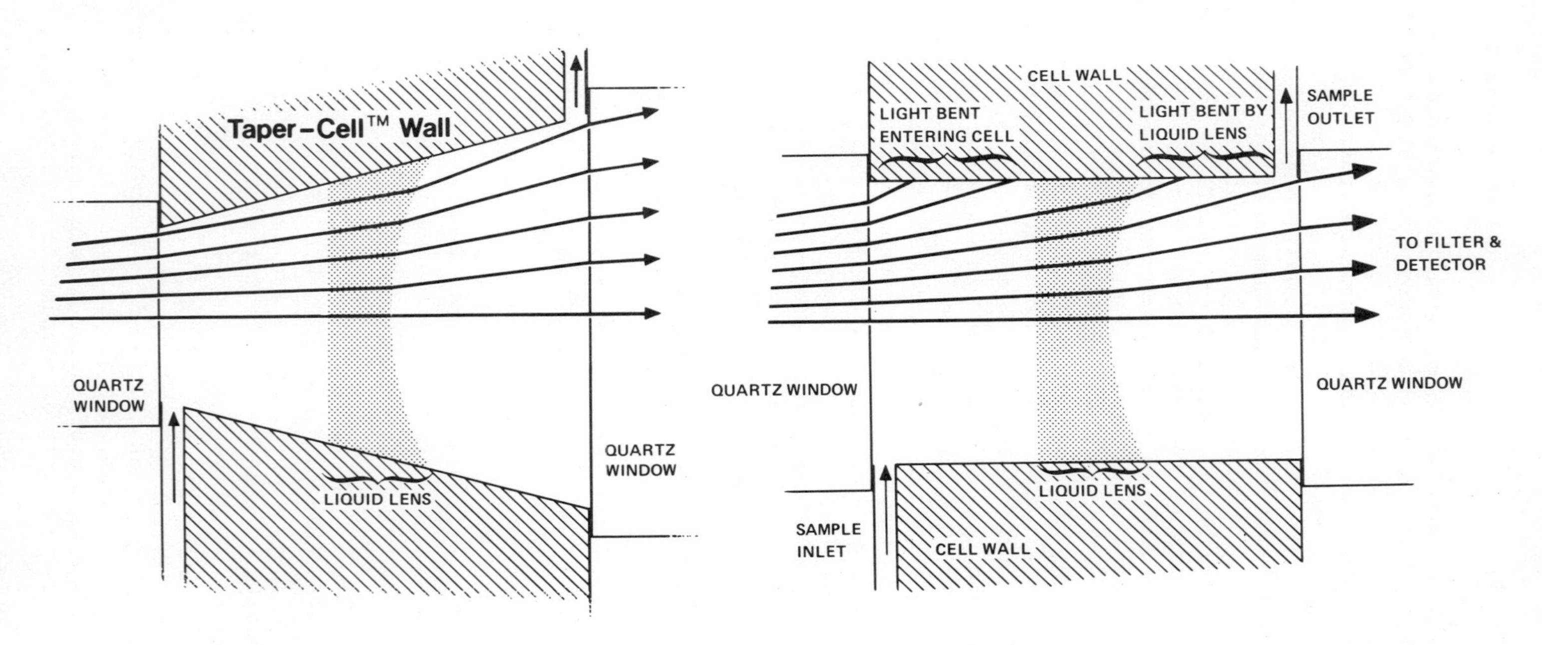

Figure 5-3 (4). Taper-cell vs. conventional flow cell.

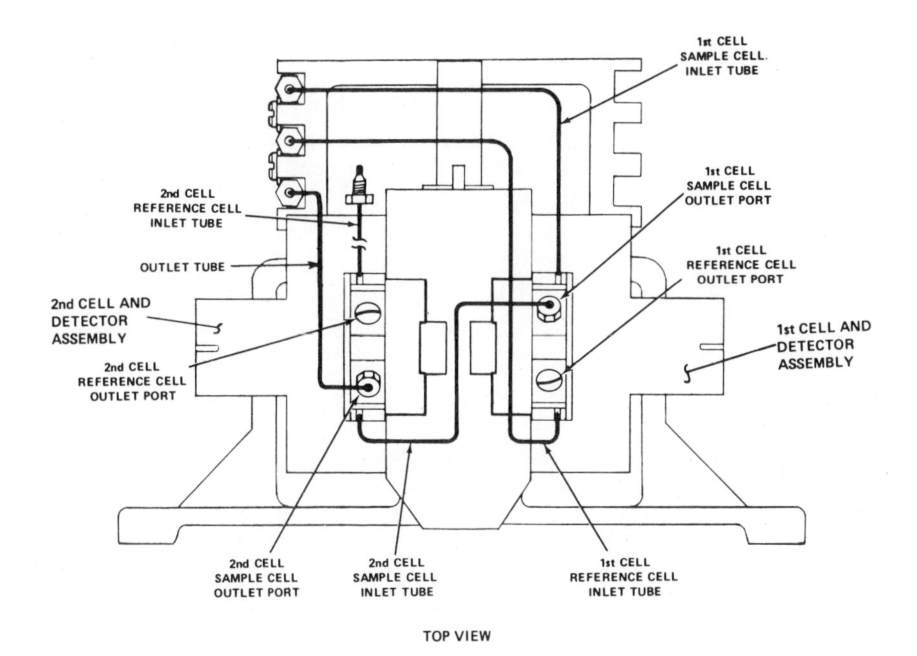

Fig. 5-4 (4). Fluid path connections for dual detector units.

wavelength monitoring at 254 nm with a fixed wavelength detector are higher than those obtainable when using multiwavelength detectors operating at 254 nm. This is because the noise level for the fixed wavelength low pressure mercury lamp and optics is considerably less than for a deuterium lamp (7).

A general rule of thumb regarding the detection limits of UV detectors at 254 nm is seen in Table 5-1. For example, at 254 nm, the detectability of the nucleic acid base, uracil, is approximately 5×10^{-10} gm./ml.

The differential refractometer detector is considered by many to be the "universal" detector. This is not necessarily true, as all of the current LC detection systems have limitations. In defining a universal detection system for LC, one must specify a type of detector that is sensitive to all classes and types of compounds, and reasonably insensitive to variations in operating parameters such as temperature, flow rate, pump (pressure) pulsations, and solvents.

TABLE 5-1

General capability of detection limits for UV detectors

Extinction Coefficient (ξ)	Detection Limit (gms.)
10^4	10^{-9}
10^3	10^{-8}
10^2	10^{-7}
10	10^{-6}

The refractometer does provide a wide range of application in that all substances have a refractive index. Sensitivity is approximately 10^3 lower than with UV detectors; however, the RI detector possesses the advantage of exhibiting "universal" sensitivity provided the RI of the solute and solvent are different. Hence, the greater this difference, then, the sensitivity for a given component is greater.

The major disadvantages of the RI detector are those of extreme sensitivity to temperature changes, flow rate fluctuations and lack of applicability for gradient elution work. Column ovens (Chapter 4) and pulseless pumps (Chapter 2) assist in minimizing the effects of temperature and flow variations.

The two designs of refractometers available are the Fresnel type and the deflection type.

Figure 5-5 is a schematic diagram of the Fresnel type refractometer detector patented by E. S. Watson (8). This refractometer uses a single-axis optical system with common optics and a single prism to illuminate the two interfaces. Watson states that inherent stability is obtained in addition to reducing problems associated with temperature fluctuations. Transmittance rather than reflectance of the interfaces is measured.

In reference to Figure 5-5, light from the source lamp SL passes through source mask M1, infrared blocking filter F, fine adjusting glass G, aperture mask M2, and is collimated by lens, L1. Mask M2 defines two collimated beams that enter the cell prism and impinge upon the two glass-liquid interfaces formed by the sample and reference liquids, which are in contact with the prism. SL and L1 are mounted in a common assembly called the projector which can be rotated about the axis of the prism. This permits coarse adjustment of

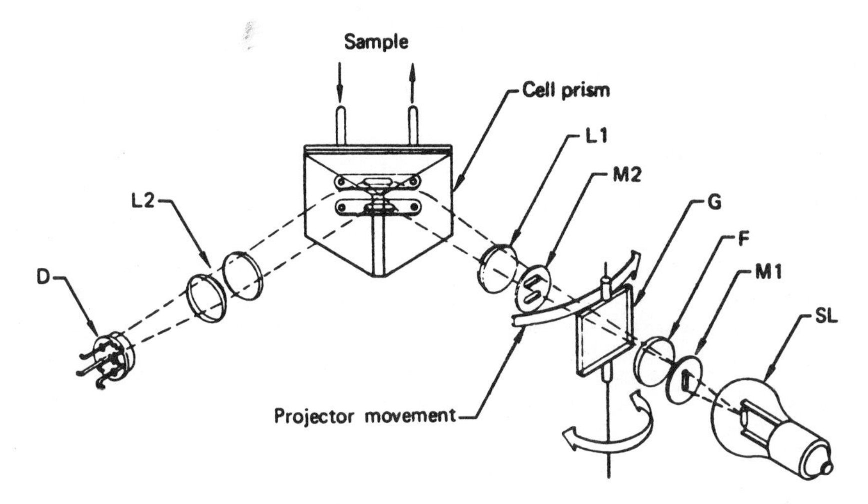

Fig. 5-5. Schematic diagram of the Fresnel type refractometer.

the incident angle to slightly less than the critical angle and is made by rotating the fine adjusting glass, G. Light from the two beams which is internally reflected does not enter detector lens L2; only that light which is transmitted through the two interfaces passes through and impinges on the stainless steel plate. The surface of this plate has a finely ground light scattering surface, and the transmitted beams appear as two spots of light. The detector lens assembly forms an image of these spots on two light sensitive elements in the photodetector, D. The photoconductor elements in the dual photodetector are arranged in a Wheatstone bridge circuit whose imbalance provides the RI measurements.

Short-term peak to noise is equivalent to 3×10^{-7} RI units. Drift rate is less than 1×10^{-6} RI units per hour. Another advantage of the Fresnel detector is that only a very thin film of liquid is required for a measurement. Fresnel equations are non-linear for parallel monochromatic rays; however, adequate linearity is obtained in the region of 10% transmittance.

Disadvantages of the Fresnel refractometer are that unstable films may form in the prism and affect measurements; additionally, these refractometers are more sensitive than deflection refractometers to bubbles and particulate matter in the cells. The Fresnel detector is also extremely flow sensitive. Flow rates above 2 milliliters per minute are virtually impossible if quantitative results are desired.

Figure 5-6 is a schematic diagram of the deflection

refractometer detector (9). This detector measures the deflection of a light beam resulting from the difference in RI between the sample and reference liquids. Deflection occurs at the surface of the cell partition permitting the use of small cell volumes. In principle, a beam of light from the lamp passes through the optical mask which confines the beam to the region of the sample cell. The mirror reflects the beam back through the sample and reference cells and through the lens coming into focus on the detector. The angle of deflection (RI between sample and reference) is determined by the location of the focused beam on the detector. As the beam moves on the detector, an output signal is generated. amplified and recorded. The optical zero glass deflects the beam from side to side to adjust for zero output signal. Advantages of deflected (transmitted and refracted) beam detection are wide-range solvent capability, minimal signal interference by bubbles or particulate matter in the solvent, the inherently high sensitivity obtained, and capability of operating at flow rates from 0.5 to 200 milliliters per minute. The latter renders this detector ideal for preparative work where high flow rates are used to maintain the desired linear velocity through the larger diameter columns.

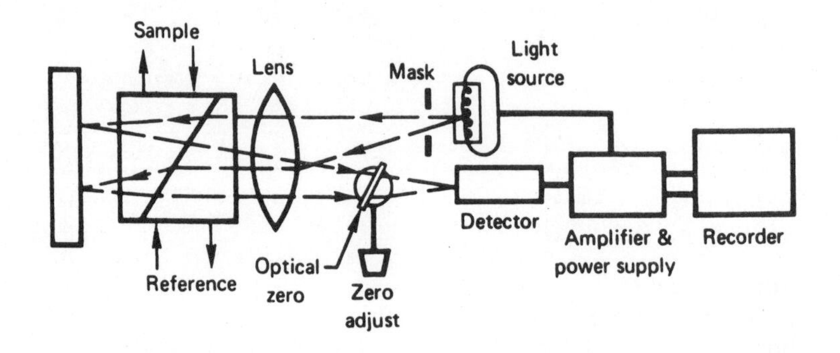

Fig. 5-6. Deflection refractometer detector. (Courtesy of Waters Associates)

In spite of some disadvantages of refractometer detectors in general, such as drift caused by temperature changes and erratic responses due to pressure fluctuations, when all parameters are optimized, changes as small as 10^{-7} RI units are claimed to be detectable.

In using the refractometer, flow troubleshooting may become important. It is often difficult to determine if the flow stream or the refractometer is causing the problem. To solve some of these problems, the following procedure may be used:

1. Stop the flow. Normally, the baseline readout will be different. Criteria such as drift due to mixing, fractionation, and cycling will virtually be eliminated. If anomalies do occur then one must suspect the detector.
2. Dual-column, matched-flow systems are recommended for determining anomalies in differential refractometer applications.
3. Non-linearity is often the problem in refractometer monitoring. Virtually the only cause for this is misalignment of the image transmitted to the photocell. This indicates that the beam does not pass through the cell properly. This may occur with either type of refractometer, Fresnel or deflection. Proper alignment is mandatory if anything other than qualitative determinations are being sought. Quantitatively, accuracy of the technique should be at least $\pm 5\%$, relative to the amount present for any given component.

The solute transport detector or flame ionization detector described by Gilding (10) involves the removal of the mobile phase prior to detection. In principle, the column eluant containing the solute to be analyzed is fed onto a moving wire or conveyor. The mobile phase is evaporated leaving the solute deposited on the wire which is subjected to a pyrolysis chamber containing a special catalyst for producing a smooth decomposition reaction. Advantages of this detector are that the mobile phase has no effect on the detector, linear response is obtained, and responds to virtually any compound containing carbon. Disadvantages are also evident in that the system is limited to non-volatile solutes, it has comparatively poor sensitivity, and it is quite bulky and expensive. Homogeneity of the coating on the moving wire for a given component is also critical; otherwise, noise and multiple peaks for a single component may be observed. The maintenance and troubleshooting regions of major concern in the solute transport detector are those of complete mobile phase evaporation, activity of the catalyst bed and reproducibility. Improvements have been made using wires coated with ceramic and metallic coatings (11); however, this is not fully developed at this time for HSLC.

The micro-adsorption detector (M.A.D.) (12), (Varian Aerograph) consists of two cell compartments; a reference cell packed with non-adsorbing material and an active cell packed with an adsorbent. A thermistor probe is mounted in each cell. When a component from a mixture is adsorbed the thermistor temperature increases due to the heat of adsorption. Upon desorption, the thermistor temperature changes, comprising the principle of the M.A.D. A Wheatstone bridge is used for measuring the temperature imbalance between the sample and reference cells.

The signal of an ideal micro-adsorption detector would be the differential of a gaussian curve. However, in practice, the signal departs from this ideal shape because of heat loss to the eluant stream and the environment. Advantages of the micro-adsorption detector are its minimal dead-volume and simplicity. The disadvantage is lack of precise temperature control as the sensitivity is an inverse function of the detector temperature. One may experience a twenty to eighty percent loss in sensitivity for a 35°C change in temperature. The activity of the mobile phase influences sensitivity also as it competes with the solute for the adsorbent packing. It may be used quantitatively; however, recent work has shown this detector non-viable for High Speed LC (13).

Fluorimeters for liquid chromatographic detection are becoming more widely used. Detection is based upon the fluorescent energy emitted from a solute excited by UV radiation. Fluorescent detectors are generally no more sensitive then conventional UV detectors; however, they do provide high selectivity. The major disadvantage associated with the fluorescent detector is its susceptibility to interference by fluorescence or quenching effects from background and non-linearity above 0.05 absorbance units.

Fluorescent detectors are generally of two types. These are either straight-through or 90-degree types. Due to stray light and subsequent light-scatter, the 90-degree type is recommended in order to minimize noise and to observe increased sensitivity. Fluorescence monitoring in tandem with UV monitoring serves many useful advantages. First of all, one may greatly simplify a complex matrix problem (inherent in UV) merely by using fluorescence. Additionally, tremendous selectivity for specific compounds may be obtained depending upon the excitation wave length of the molecules in question. The availability of dialing a particular excitation wave length has great advantage over changing filters for the same purpose. There are many compounds, UV sensitive, which do not possess fluorescent qualities. In biochemistry and pharmaceutical chemistry, sometimes the metabolites of an administered drug will possess fluorescent properties. Therefore, in practice, fluorescence combined with UV has the capability of providing a lot of quantitative data on metabolic processes among various subjects. Key areas are proper solvent choice and good calibration standards. These standards will provide good linear relationships in the low ppm to nanogram level. As the concentration of component increases, it will become necessary to recalibrate and construct a new working curve because quenching effects, sample concentration, and system anomalies deem it necessary. This is especially true in going from sub-nanogram to microgram quantities.

LC detectors have many sources of noise. The four major causes of noise are those associated with the electro-optical system, temperature variations, fluctuations in chemical composition, and flow rate and pressure fluctuations (14). High-frequency noise is defined as greater than one cycle per five seconds. This noise generally has no effect on the detection limits because of its high frequency and low amplitude. Short-term noise is within the range of one cycle per five seconds to one cycle per five minutes. It is the limiting noise for fast and medium speed peaks. Drift is a baseline shift over a period greater than five minutes. The major factor contributing to this drift is ambient temperature change. This noise does not affect the detection limit of slow eluting components.

Temperature fluctuations, as mentioned earlier, cause baseline drift; however, with UV detectors this is minimal compared to differential refractometers. If the detector is housed in a well-ventilated compartment (fan, vents, etc.) the problem is minimized. Fluctuations resulting from variations in the chemical composition of UV absorbers in the solvents being used will produce large baseline irregularities. Table 5-2 (14) shows the relative UV absorbance of common solvents listed in a eluotropic order. The length of the horizontal line from each solvent depicts the regions where it cannot be used in UV detection.

Bubbles resulting from improper system purging, leaks, and solvents which have not been degassed, present the user with unnecessary problems and contribute to appreciable downtime. As mentioned earlier, these bubbles may be observed either on the bridge excitation meter or the recorder trace upon passing through the detector. Corrective measures are pre-purging the column, tightening and/or replacement of leaking fittings or septa, and proper degassing of solvents. Problems may occur, however, when attempting to dislodge an entrapped bubble in any "larger-than-necessary" dead-volume area. This occurs quite frequently in aqueous systems and in high vapor pressure organic systems.

Several techniques may be used depending upon location of bubble problems. Some workers have found that placing back-pressure on the system will displace the lodged bubbles. This is accomplished by *momentarily* fitting a plug (septum or silicone rubber) onto the end of the detector exit or exit tube during operation. The subsequent pressure build-up will often free the system of bubbles. This technique must be

TABLE 5-2

Relative UV absorbance of solvents (14)

Solvent	R I	UV cutoff 200nm 250 300
n Pentane	1.36	
Petroleum ether	–	
Cyclohexane	1.43	
Carbon tetrachloride	1.47	
Amyl chloride	1.41	
Xylene	1.50	
Toluene	1.50	
n-Propyl chloride	1.39	
Benzene	1.50	
Ethyl ether	1.35	
Chloroform	1.44	
Methylene chloride	1.42	
Tetrahydrofuran	1.41	
Ethylene dichloride	1.45	
Methylethylketone	1.38	
Acetone	1.36	
Dioxane	1.42	
Amyl alcohol	1.41	
Diethylamine	1.39	
Acetonitrile	1.34	
Pyridine	1.51	
n-propanol	1.38	
Ethanol	1.36	
Methanol	1.33	
Ethylene glycol	1.43	
Acetic acid	1.37	
Water	1.33	
		254

used with care as the detector may become damaged. Increasing the pressure delivered by the pump will often be sufficient to dislodge the entrapped bubble, especially in the vicinity of the injector and pre-column plumbing. However, most problems seem to occur in the vicinity of the detector, either within or immediately before entering the detector. During normal operation, the pressure-drop along the column is quite large resulting in only minimal pressure being imposed at the entrance to the detector. Entrapped bubbles may be successfully purged-out of the system by attaching a two- or three-foot section of small inside diameter stainless steel or polyethylene tubing to the cell exit and elevating this above the detector maintaining a small but constant back-pressure on the flow cell.

REFERENCES

1. S. H. Byrne, Jr. in *Modern Practice of Liquid Chromatography*, J. J. Kirkland, Ed., John Wiley and Sons, Inc., New York, 1971

2. Laboratory Data Control *Duo-Monitor Bulletin*, 1970, LDC, a Division of Milton Roy Company.

3. *Chromatronix Lab Notes*, 1972, Spectra Physics Corp.

4. Waters Associates, Private Communication, 1974.

5. Varian Instruments, Laboratory Notes, 1972.

6. Schoeffel Instrument Corporation.

7. A Sonnenschein "Analytical Instrumentation" Vol. 12, ISA AID 74425 Instrument Society of America, 20th Annual Mtg., May 1974.

8. E. S. Watson, *American Laboratory*, October 1969.

9. Waters Associates Technical Laboratory Bulletins.

10. D. K. Gilding, *American Laboratory*, October 1969.

11. V. Pretorius and J. vanRensburg, *J. of Chrom. Sci.*, Vol. 11, p. 355, July 1973.

12. F. R. MacDonald, C. A. Burtis, and Jack M. Gill, *Research Notes*, Varian Aerograph, July 1969.

13. R. P. W. Scott, *J. of Chrom. Sci.*, Vol. 11, p. 349, July 1973.

14. *Lab Notes 4*, Chromatronix, Div. of Spectra Physics, Jan. 1971.

QUESTIONS AND ANSWERS

Question: Sometimes when using a UV detector, why are negative peaks observed?

Answer: This occurs when non-UV absorbing constituents pass through the cell. Actually, a change in refractive index is being observed due to light scattering phenomena.

Question: Sometimes the major component in the mixture being analyzed reaches a "peaked-out" null point at about 60-70% of full scale on the recorder trace when using an attenuation of X 256 with a UV monitor. What causes this?

Answer: This is due to saturation of the detector electronics caused by too high a concentration of the sample component. Use a more dilute mixture and keep in mind that generally at or slightly above attenuations of 1.28 AUFS most UV detectors become non-linear.

Question: Why do both positive and negative peaks appear when using refractive index detectors?

Answer: This indicates refractive index differences between the solvent (mobile phase) and the solute(s) being analyzed. Conceivably, those solutes with RI greater than the mobile phase in the reference cell will produce positive peaks whereas, those with RI less than the mobile phase will produce a negative response. Therefore, when selecting a mobile phase for RI monitoring, choose one which reflects the maximum in RI difference between itself and the solutes in the mixture. This, accordingly, will yield higher sensitivity.

Question: Why, upon selecting two solvents with approximately the same RI for gradient work, is a random, drifting baseline obtained during the gradient?

Answer: There several reasons why this occurs, all of which seem to nullify the RI detector for gradient monitoring. The more common causes are as follows:

(a) The RI detector is very sensitive to even minute changes in flow. This difference will be critical in both the sample and reference cells.

(b) The RI detector is sensitive to very small temperature changes which do, in fact, occur quite readily when two solvents are being continually mixed.

(c) Two solvents, independently having the same refractive index, may exhibit different refractive indices while being continually mixed.

Question: Having optimized the mobile phase and solute to reflect the maximum in ΔRI the sensitivity obtained during analysis is still inadequate? Why?

Answer: This sounds as if there were an optical alignment problem or possibly a defective lamp.

Chapter 6

COMPREHENSIVE TROUBLESHOOTING OF LC SYSTEMS

As most analytical problems in liquid chromatography are approached using the "sequential analysis" or "first-things-first" method, virtually the same approach may be employed in developing a comprehensive system for trouble-shooting of the LC equipment. In liquid chromatography possible "trouble areas" are quite numerous and may be associated with any one major part of the system or a multiplicity of several malfunctions, in combination, from more than one major part of the system. Major parts of the system are: the solvent transport system, sample introduction, column, detector and recorder.

A comprehensive troubleshooting guide should provide a method for rapidly pinpointing the symptoms and causes associated with a particular malfunction. Additionally, it should provide guideline remedies and corrective procedures for restoring the equipment to its maximum capability.

Management, upon spending thousands of dollars for equipment and associated apparatus, needs the assurance that its investments are being utilized to the utmost. This places an added and well-justified responsibility upon all who design, sell, purchase, and use liquid chromatographic equipment. Recognizing that no system, chromatographic or otherwise, is exempt from maintenance and troubleshooting problems, it prompts us, as buyers and users, to make wise choices in our purchases and also maintain the optimum performance from our equipment.

Preventive maintenance is a milestone in achieving maximum performance. A properly executed preventive maintenance program will save costly downtime. A preventive maintenance program cannot provide nor guarantee trouble-free operation. Therefore, when anomalies inadvertently occur, we should be prepared to diagnose the problem rapidly and either remedy or provide guidance to the subsequent restoration of the equipment to its optimum capability.

The ultimate solution for accomplishing this is either via training and experience or a readily accessible guide which may partially substitute for the lack of training or experience. The following is such a guide which may provide rapid troubleshooting tips for the experienced chromatographer as well as the novice in the field. The general format of the guide will be a description of *symptoms, causes* and *corrective actions* for each major category of the liquid chromatographic system. These are:

I. *Solvent Transport System*
Solvent degassers, solvent reservoirs, pumps, and associated transfer lines.

II. *Sample Injection System*
Septum injectors and loop injectors. Solvent leaching of septa, leaking of septa and loop injectors, particulate matter.

III. *Columns*
Column bleed, leaking fittings, column plugging.

IV. *Detectors*
Low sensitivity, non-linearity, noise, erratic response.

V. *Recorders*
Malfunctions diagnose from recorder traces, recorder malfunctions, noise, electrical connections.

In conclusion, the troubleshooting tips outlined in this chapter were obtained from several sources (1, 2, 3) as well as our own experiences and are representative of most malfunctions which may occur in liquid chromatography.

Symptom	Cause	Corrective Action
Noisy Baseline (or)	A. Detector	A. Detector
	1. Contamination in sample or reference cells.	1. Flush sample and reference cells with fresh solvent.
	2. Bubbles in sample cells or reference cell.	2. Increase flow rate or place restriction at detector exit.
	3. Defective UV lamp.	3. Replace UV lamp.
	B. Solvent Transport System	B. Solvent Transport System
	1. Bubbles passing through sample or reference cells. Solvents not adequately degassed.	1. Degas solvent.
	2. Pulses from pump stroke.	2. Incorporate pulse dampener or reduce pump stroke.
	C. Column	C. Column
	1. Particulate matter from column passing through or into detector.	1. Check column exit.
	2. Leaking fitting or connector.	2. Tighten or replace fittings.
	D. Sample Introduction System	D. Sample Introduction System
	1. Leaking septum.	1. Replace septum.
	2. Partial blockage in loop injector due to particulate matter.	2. Clean loop injector
	E. Recorder	E. Recorder
	1. Grounding problem in recorder or instrument.	1. Check all recorder and instrument grounds. Eliminate ground-loops.

Symptom	Cause	Corrective Action
Drifting Baseline (short-term and long-term drift)	A. Detector 1. Contamination in sample or reference cells 2. Changes in temperature of detector. 3. Contamination build-up in cell. 4. Dust or condensate contamination in optical system. 5. Weak source lamp.	A. Detector 1. Flush cells with solvent. 2. Control temperature using constant temperature bath or dual-column arrangement. 3. Clean cell using 6N nitric acid and distilled water. 4. Wipe clean with lint-free tissue or cloth. 5. Replace source lamp.
	B. Solvent Transport System 1. Bubbles in mobile phase. 2. Contaminant in solvent reservoirs slowly being dissolved into mobile phase. 3. Previous solvent not completely removed. 4. Solvent demixing (or) nonhomogeneity.	B. Solvent Transport System 1. Degas mobile phase. 2. Wash reservoirs and replace old solvent with freshly prepared one. 3. Allow adequate purge-time. 4. Insure solvent compatability.
	C. Column 1. Column bleed.	C. Column 1. Saturate mobile phase with stationary phase or use a pre-column.

Symptom	Cause	Corrective Action
Drifting Baseline	C. Column (con't)	C. Column (con't)
	2. Strongly adsorbed component(s) being eluted from column.	2. Elute components from column. Use elustropic solvent series when necessary. Select mobile phase with greater compatability for all components of mixture, or increase temperature.
(short-term and long-term drift)	D. Sample Introduction System	D. Sample Introduction System
	1. Septum leaching caused by septum/mobile phase in-compatabiliby.	1. Select recommended septum for use with various mobile phases.
	2. Plugging of injector by particulate matter from solvents or septa causing pressure changes.	2. Clean sample introduction system.
	D. Recorder	E. Recorder
	1. Weak tubes in amplifier.	1. Check tubes and replace if necessary.

Symptom	Cause	Corrective Action
Baseline "Stair-stepping" and peaks are "Flat-topped." Baseline does not return to zero. 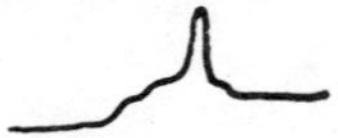	A. Recorder 1. Gain and damping control not properly adjusted. 2. Recorder or instrument not properly grounded.	A. Recorder 1. Adjust gain and damping controls. 2. Properly ground via a true earth ground.

Symptom	Cause	Corrective
"Spiking" on recorder trace.	A. Detector and Solvent Transport 1. Bubbles passing through detector (or) bubble lodged at detector entrance.	A. Detector and Solvent Transport System 1. Degas eluants and purge entire system adequately.
	B. Other Systems 1. Externally located electrical systems "tapped" off a common electrical supply, i.e., other chromatographs, thermostated oven, etc. causing intermittent line-voltage fluctuations feeding-back to the LC system.	B. Other Systems 1. Eliminate possible line voltage feedback fluctuations from other systems. Check spikes for time regularity and magnitude (duration).

Symptom	Cause	Corrective Action
Negative peaks on recorder trace	A. Detector	A. Detector
	1. Detector output polarity.	1. Change polarity, + or − , of detector output switch.
	2. In using RI detectors, some components in the mixture may have RI greater or less than mobile phase.	2. Select mobile phase of greater or less RI, (or) set detector and recorder "zero" at mid-scale initially.
(or)	3. Sometimes observed when using UV detectors and non-UV absorbers pass through detectors. It is actually an RI measurement due to light-scattering phenomena.	3. No corrective action.
	B. Solvent Transport System	B. Solvent Transport System
	1. Large quantities of air in mobile phase.	1. Degas mobile phase.
	C. Column	C. Column
	1. Air pocket displaced from column.	1. Adequately purge column with mobile phase.
	D. Sample Introduction System	D. Sample Introduction System
	1. Negative peak upon injecting sample.	1. Install loop injector.
	2. Septum begins leaking upon injection.	2. Replace septum.
	3. Injection of sufficient quantities of air.	3. Displace air from syringe before injection. Purge sample loops free of air before introducing into the air system.

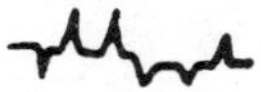

Symptom	Cause	Corrective Action
Poor Peak Shape	A. Column	A. Column
	1. Column Overload.	1. Reduce amount of sample. For *GPC*, use apprx. 0.25% solution of sample/solvent. For *LLC* and *LSC*, use apprx. 0.5 to 0.10% solution of sample/solvent. For *IEC* using pellicular ion exchange resins, anything in excess of 50 to 100 μg of any component may result in column overload.
	2. Adsorption of sample on ion exchange (IEC) column.	2. Lower sample concentration or increase the ionic strength of the eluant. Altering the pH of the eluant will have marked effects also. Increasing the column temperature will subsequently increase the ion exchange process and reduce mobile phase viscosity.

Symptom	Cause	Corrective Action
Loss of Resolution	A. Column	A. Column
	1. Loss in column efficiency and selectivity.	1. Replace column or regenerate column.
	2. In LLC, loss of liquid phase from column.	2. Replace column and use a mobile phase that does not remove liquid phase. Also, permanently bonded liquid phases should be considered, i.e., Bondapak (Waters), Permaphase (DuPont), whereby removal of liquid phase by the mobile phase cannot occur.
	3. Column overload.	3. Lower sample concentration.
	4. Increase in column flow rate.	4. Decrease flow rate.

Symptom	Cause	Corrective Action
Increased Retention Volumes 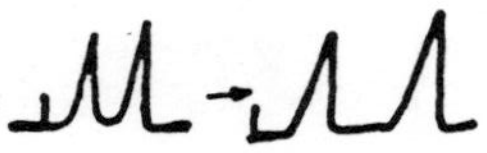	A. Column 1. In LSC, activity of column is increasing . Solvent is stripping water from column. 2. In LLC, liquid phase has been lost from the column. 3. Temperature of column is too low.	A. Column 1. Add water to mobile phase or replace with new column which has been properly deactivated. 2. Replace column and change mobile phase. 3. Increase column temperature.
	B. Solvent Transport System 1. Flow rate of mobile phase is too low. 2. In IEC, ionic strength of mobile phase is too low. 3. In IEC, pH may be too high or too low.	B. Solvent Transport System 1. Increase flow rate. 2. Increase ionic strength of mobile phase to shorten retention volume. 3. Alter pH; this may alter the elution order also.

Symptom	Cause	Corrective Action
Recorder will not "zero."	A. Recorder 1. Power to pen not turned "ON." 2. Recorder output "dead."	A. Recorder 1. Check power switch. If already turned "ON," switch may be defective. 2. Check recorder "ZERO" by shorting the input from the detector, i.e., turn attenuator switch to infinity ( ) and adjust recorder zero. If this is possible, problem is probably associated with the input signal from the detector. If, in this position, the recorder will not zero, the problem is associated with the recorder.
	B. Detector 1. Poorly connected electrical leads from detector output to recorder terminals.	B. Detector 1. Check and secure electrical leads from detector to recorder.

(Cont'd.)

Symptom	Cause	Corrective Action
Recorder will not "zero."	2. Bubble in sample cell or reference cell.	2. Increase flow rate to purge out bubble or place restrictor at detector exit to displace bubbles.
	3. Physical blocking of sample or reference image from detector.	3. Check light path and remove obstructions.
	4. Source lamp defective.	4. Replace source lamp.
	C. Column	C. Column
	1. Excessive column "bleed."	1. Use balancing dual-column technique.
	2. Air being displaced from column.	2. Adequately purge column with mobile phase.
	3. Previous mobile phase not removed.	3. Purge column free of previously used eluants.
	D. Solvent Transport System	D. Solvent Transport System
	1. Previous solvent not removed.	1. Allow pump and associated transfer lines to be completely purged free of previous eluants.

Symptom	Cause	Correction Action
No flow through column and no pump pressure	A. Solvent Transport System	A. Solvent Transport System
	1. Solvent reservoir empty.	1. Refill reservoir with desired solvent.
	2. Air in pump.	2. Siphon liquid into pump and resume pump action. Prime pump.

Symptom	Cause	Corrective Action
Pump exerting pressure on system, but no flow through column	A. Solvent Transport System	A. Solvent Transport System
	1. Flow restriction in pump or in pump transfer lines.	1. Check septum, relief valve (defective?), transfer line from pump to injector. Check all associated fittings and couplings from pump to detector exit.
	2. Restriction to flow in the vicinity of the pump, injector or column exit to the detector.	2. Particulate matter will clog the small diameter lines to the pulse dampeners on pulsating pumps; therefore, isolate these before encountering downstream pursuits.
	B. Sample Introduction System	B. Sample Introduction System
	1. Plugging in the sample in inlet due to particulate matter from septa, samples, etc.	1. Filter all solvents, even distilled H_2O.
	2. Syringes.	2. Exercise care with syringes. Use syringes that are free of barbs, etc., which may tear rather than pierce the septum upon sample injection.
	C. Column	C. Column
	1. Leaks in fittings.	1. Check all fittings for leaks.
	D. Detector	D. Detector
	1. Leaks	1. Check detector and associated fittings.

REFERENCES

1. N. Hadden and F. Zamaroni, *Troubleshooting and Maintenance,* Varian Associates, 1970.

2. L. R. Snyder and J. J. Kirkland, *Modern Liquid Chromatography* (ACS) Short Course), American Chemical Society, 1971.

3. *Instruction Manual* (98526), Waters Associates, Inc., 1970.

Part II

GAS CHROMATOGRAPHY

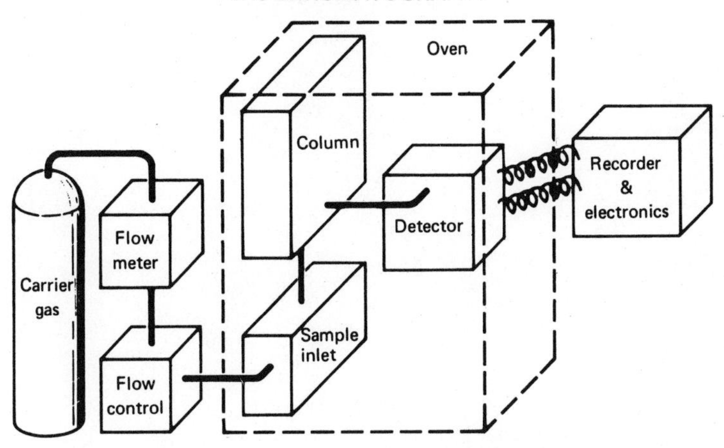

PART II. GAS CHROMATOGRAPHY

This part of the text is concerned with Gas Chromatography (GC) and begins with the basic fundamentals. Following this brief introduction chapter, there are practical approaches to understanding and maintaining the following GC systems: the Carrier Gas-Inlet, the Column-Oven, the Detector, the Recorder and Electronics. A final chapter on Comprehensive Trouble Shooting, is a ready reference for fast GC problem solving.

One section of Part II includes maintenance and problems with chromatographic integral electronics (i.e., power supplies, electrometers, recorders, and integrators). Another section includes maintenance of auxiliary equipment interfaces (i.e., ultraviolet, infrared and mass spectrometers). A trouble-shooting section concerning thermal degradation (pyrolysis) inlet systems is also included in Part II.

Chapter 7

INTRODUCTION TO GAS CHROMATOGRAPHY

Gas chromatography (GC) is an instrumental method of analysis for the separation, identification, and quantitation of volatile mixtures. "Permanent" gases (such as oxygen and carbon dioxide), volatile liquids, and pyrolyzed solids can all be separated by gas chromatographic techniques. Part II of the text will be concerned primarily with the GC instrumentation which permits these analyses to be carried out. Some basic knowledge of the principles of separation is necessary, however, for a full appreciation of the material presented in subsequent chapters. A detailed explanation of the theory will not be necessary. For a more rigorous development of the theory, texts by Keulemans (1), Zlatkis (2), and others should be consulted.

This first chapter, dealing with the basic fundamentals of gas chromatography, includes some important definitions and formulas which are used to describe the performance of GC instruments. In later chapters the characteristics, the maintenance, and the troubleshooting of various segments of the gas chromatograph will be discussed separately including inlet systems, columns, detectors, and electronics. Finally, a summary chapter is included which is a comprehensive guide for determining the causes and remedies for problems occurring in the GC.

THE PRINCIPLE

Separation is the primary function of a gas chromatograph. This process of separation takes place inside the column, a length of tubing containing a packing material which consists of a stationary phase coated on an inert support. As illustrated in Figure 7-1, the mixture, consisting of components A and B, is introduced into a gaseous mobile phase which is constantly traversing the length of the column. While moving through the column, components A and B will interact with the stationary phase while the mobile phase does not interact at all. For this reason, A and B are delayed as they move through the column. In Figure 7-1, component B interacts with the stationary phase to a greater extent than component A, and, consequently, the mixture begins to separate. Ideally, the mixture is completely separated by the time the components reach the end of the column.

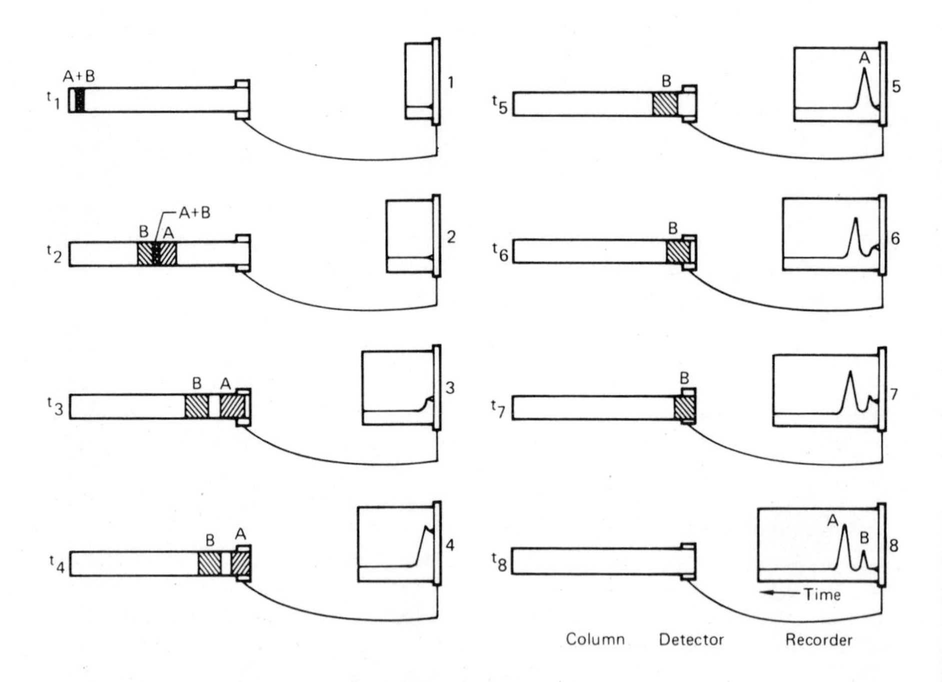

Fig. 7-1. The operation of a chromatograph is shown in its most basic concept. A mixture of components A and B enters the separation column, at left, where it is separated into pure A and pure B. The arrival of the components at the end of the column is observed by the detector which signals a strip chart recorder. The end result is a chromatogram which is representative of the original mixture. (Courtesy of "Chemistry" and American Chemical Society) (3).

The elution of the components from the column is detected electronically by a suitable detector, and signals are sent to a strip chart recorder which produces the chromatogram. This chromatogram consists of a series of peaks, each of which indicates the elution of a component and the amount of component present. The time of elution may be used to identify the components of the mixture and is defined as the elution time or retention time for a particular component.

A basic gas chromatograph, consisting of the following systems, is shown in Figure 7-2:

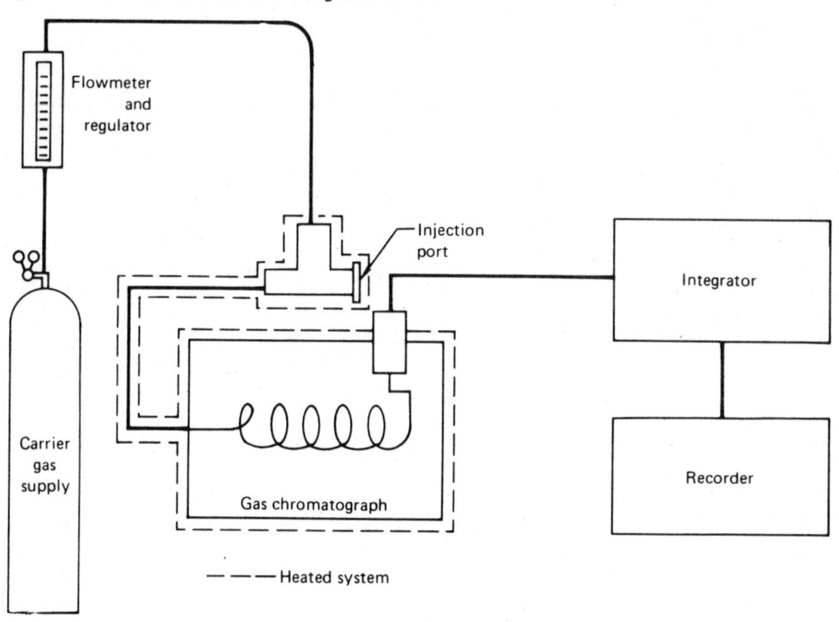

Fig. 7-2. The basic parts of a GC system are shown with the heated areas surrounded by broken lines. An integrator is sometimes employed to measure the area under the peaks of the chromatogram. (4)

In an ideal situation, the peaks of the chromatograms should be completely separated. The peak width at the base should be narrow, even at the end of the chromatogram. These performance factors are measured in terms of *efficiency* and *resolution* which can be calculated from a few simple measurements made on the chromatogram. A typical chromatogram is shown in Figure 7-3. The time required for the peak to appear, called the retention time (t_R), measured from the point of sample injection to the point at which the peak is at its height. The width of the peak (W) is measured at the bottom of the peak (base-line) between lines drawn

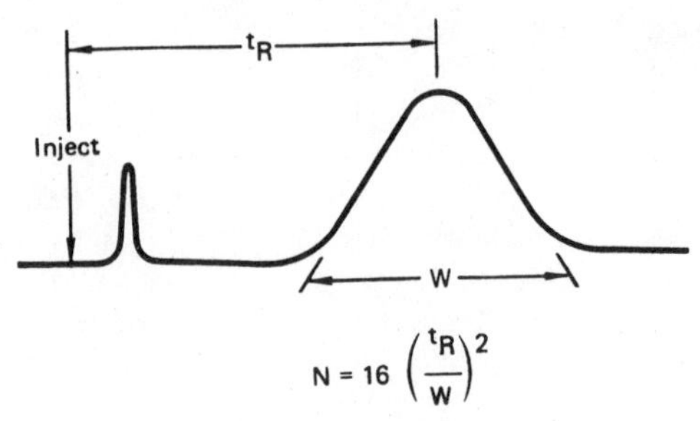

$$N = 16\left(\frac{t_R}{W}\right)^2$$

Fig. 7-3. A chromatogram consists of a set of peaks produced by the chart recorder. The retention time of a peak (t_R) is measured from the point of injection to the highest point of the peak. Peak width (W) is measured at the base of the peak between lines drawn tangent to the sides of the peak. From these two dimensions, column efficiency (N) can be calculated. (Courtesy of Varian Aerograph) (5)

tangent to the sides of the peak shown in Figure 7-3.

Column efficiency, given by the *number of theoretical plates* (N) is calculated from the expression,

$$N = 16\left(\frac{t_R}{w_b}\right)^2 \quad \text{(approximate value}$$

where,

t_R = retention time (here measured in units of length).

w_b = peak width at the base.

N = number of theoretical plates.

Efficiency is a function of column length among other parameters; and, for purposes of comparison, efficiency is commonly expressed in terms of *Height Equivalent to a Theoretical Plate* (HETP). HETP is the column length (L) divided by the number of theoretical plates (N), as follows:

$$\text{HETP} = L/N$$

A third expression for column efficiency is the reciprocal of HETP which expresses efficiency in terms of plates per foot. All three expressions for efficiency are used.

The ability of a column to separate a given pair of compounds can be measured in terms of resolution (R). As illustrated in Figure 7-4, resolution is a function of peak width and distance between peaks.

I. Pneumatic and Sample Systems (Chapter 8)

A. Carrier Gas (mobile phase) - Inert gas used to move the sample through the column.

B. Pressure and/or Flow Control Apparatus - Maintains a constant pressure and/or flow rate of carrier gas through the column. It may also vary column pressure at a predetermined rate.

C. Sample Port (injector) - for introduction and vaporization of the sample.

D. Pyrolyzers - for the thermal degradation of solid and liquid samples prior to introduction to the column. (Chapter 9).

II. Columns and Column Ovens (Chapter 10).

A. Separating Column - tubing containing the stationary phase which is coated either on an inert solid support or on column walls.

B. Ovens, Heaters, and Controllers - controls the temperature of the column, detector, and injector.

III. Detectors and Related Electronics (Chapter 11)

A. Detector-Detects sample components as they elute from the column and provides the basis for quantitative measurement. (Chapter 11).

B. Power Supply and/or Electrometer Circuits - Amplifies the detector signal which is sent to the recorder and provides the detector with needed power (Chapter 12).

C. Recorder - Provides a permanent visual record of the analysis (Chapter 12).

IV. Auxiliary Systems (Chapter 13).

A. Mass Spectrometers, infrared spectrophotometers, and ultra-violet spectrophotometers aid in the positive identification of the sample.

B. Mechanical and digital integrators aid in quantitating chromatographic peaks (Chapter 12).

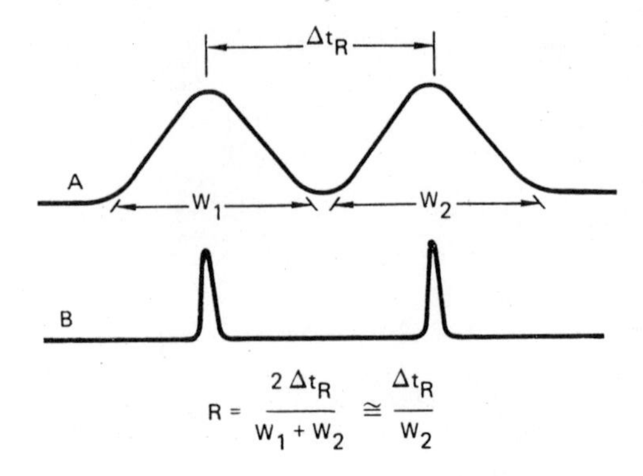

$$R = \frac{2\,\Delta t_R}{W_1 + W_2} \cong \frac{\Delta t_R}{W_2}$$

Fig. 7-4. The ability of a particular column to separate a given pair of compounds can be measured in terms of resolution (R). As illustrated, R is a function of peak width and the distance between two peaks. Usually R is measured for the pair of peaks which are the most difficult to separate. (Courtesy Varian Aerograph) (5)

R = column resolution
W_1, W_2 = peak width at base
Δt_R = distance between the apex of the two peaks

A very useful equation for comparing column dimensions and analysis conditions was proposed in 1956 by van Deemter (3). The simplest form of the equation, relating HETP and carrier gas velocity, is written as follows:

$$\text{HETP} = A + \frac{B}{\mu} + C_{\mu}$$

This equation can be plotted with HETP as a function of carrier gas velocity (μ) as shown in Figure 7-5. The curve has a minimum of a particular value of μ called the optimum carrier gas velocity (μ_{opt}). At this point HETP is at its lowest value and the column efficiency is, thus, at its highest.

Van Deemter plots can also be used to compare various types of columns and stationary phases (6). The effect of HETP on the amount of liquid stationary phase used in gas-liquid chromatography is shown in Figure 7-6. The percent liquid phase is a relative figure representing the percentage of liquid phase in the solution used to coat the column or column support. Columns with a lower liquid load exhibit lower minimums of HETP and are thus capable of higher efficiency.

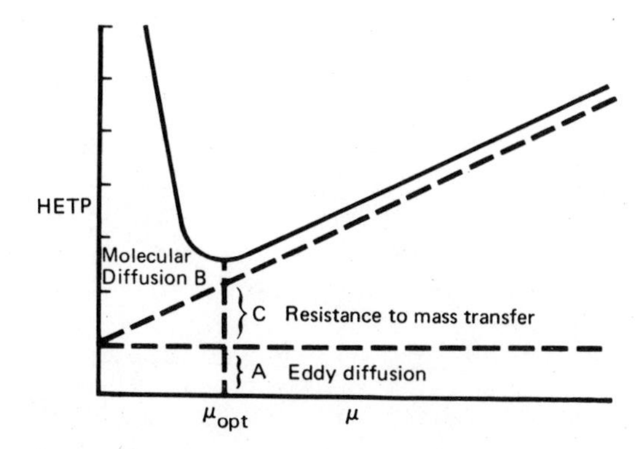

Fig. 7-5. Van Deemter proposed an equation relating HETP to carrier gas velocity. A, B, and C are constants which describe the physical characteristics of the column affecting HETP. A plot of HETP versus carrier gas velocity will reveal a curve with a minimum at a particular carrier gas velocity which should provide the highest column efficiency. (6)

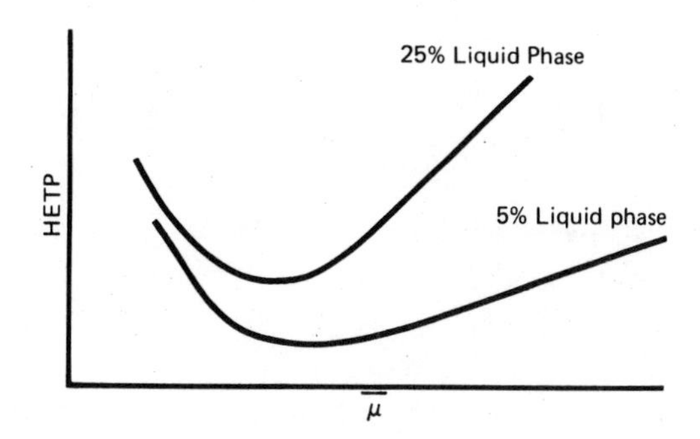

Fig. 7-6. Lower amounts of liquid phase (determined by the percentage of liquid phase in the solution which was used to coat the column support) will provide more efficient columns. (5)

HETP is dependent upon the type of carrier gas used. Some carrier gases are able to diffuse more readily through the solid support. They are, as a result, better able to move through the column. Relative van Deemter plots for nitrogen, having a higher diffusivity, displays a lower minimum HETP (Figure 7-7).

Column diameter affects the van Deemter plot as shown in Figure 7-8. Reducing the internal diameter of a column, all other factors being held constant, will result in a more

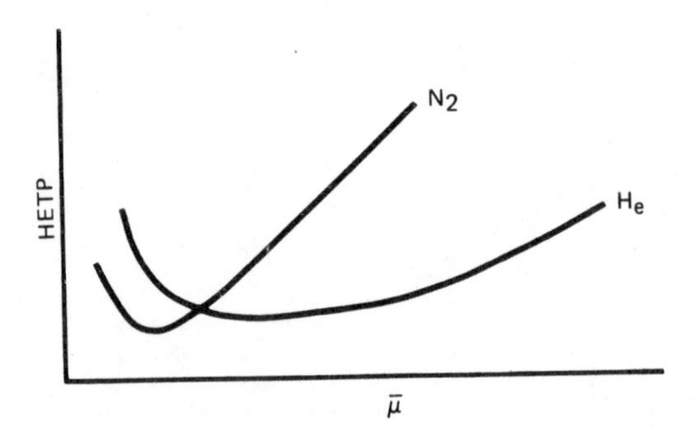

Fig. 7-7. Different carrier gases exhibit different HETP plots. Helium carrier gas has a wider range of good carrier gas flows, however nitrogen is capable of higher efficiency. (5)

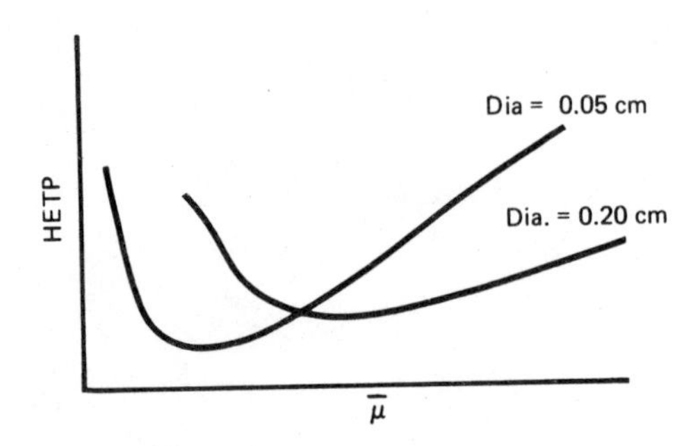

Fig. 7-8. Smaller diameter columns are more efficient but require smaller samples. The minimum practical column diameter (due to packing difficulties) is about 0.03 cm. (5)

efficient column. Difficulties may occur when packing the smaller diameter columns, but columns with internal diameters as low as .025 cm are used frequently in situations where high resolution is required. The small diameter columns also require the use of comparatively smaller sample sizes.

If too much sample is introduced into the column, flooding will result causing a loss in efficiency. As seen in Figure 7-9, as sample size increases for a given column size, column efficiency remains somewhat constant up to a point, after which column efficiency begins to drop rapidly.

Very often, in formulating a chromatographic system, highest efficiency and resolution are not the only factors

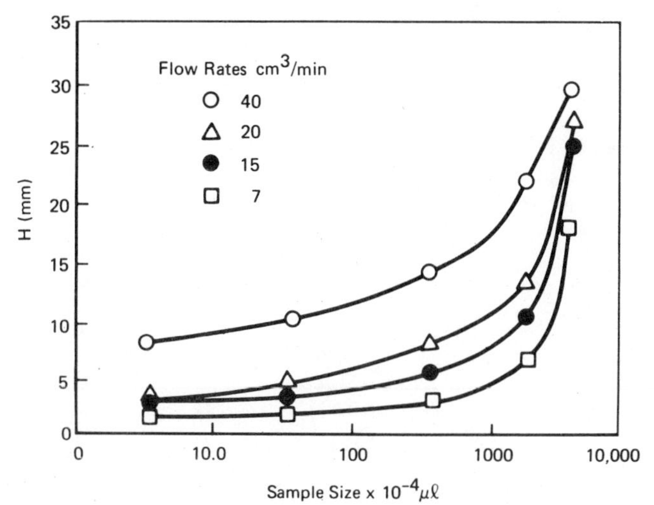

Fig. 7-9. Sample size is restricted by column size and column type (packed or open tubular). As sample size increases for a given column, efficiency remains relatively constant up to a point at which efficiency drops rapidly. (Courtesy J. Gas Chromatog.*)* (7)

to be considered. High resolution gas chromatography can be slow and requires smaller sample sizes. These factors may not be compatible with the analyst's needs. Therefore, the chromatographer may sacrifice some resolution by increasing the flow rate in order to speed the analysis. He may choose to use a larger diameter column in order to increase his sample size. Low liquid loaded columns do provide the best resolution, but, unfortunately they can deteriorate rapidly, particularly if large volume samples are used. The needs and requirements of each analysis must be considered in designing the best chromatographic system.

With just these few terms and quantities, plus a few more discussed later, an understanding of the individual GC systems can be readily accomplished. These quantities will be referred to several times in later chapters, and it is recommended that they be understood fully before continuing.

REFERENCES

1. L. S. Ettre and A. Zlatkis, Ed., *The Practice of Gas Chromatography*, Wiley, N.Y., (1968).

2. A. I. M. Keulemans, *Gas Chromatography*, 2nd ed., Reinhold, New York, (1959).

3. W. R. Supina and R. S. Henley, *Chemistry*, *37*, 12 (1964).

4. Private communication, M. T. Jackson, Jr.

5. F. Bauman and J. M. Gill, Aerograph Res. Notes, (1966).

6. J. J. VanDeemter, F. J. Zuiderweg and A. Klinkerberg, *Chem. Eng. Sci.*, *5*, 271 (1956).

7. A. Zlatkis and J. Q. Walker, *J. Gas Chromatography*, *1*, 10 (1963).

Chapter 8

PNEUMATIC AND SAMPLE INTRODUCTION SYSTEMS

Routine maintenance and operating principles of the pneumatic system and the sample injection system of a gas chromatograph are discussed in this chapter. Malfunctions in the pneumatic injection system can mask problems in other systems of the chromatograph; thus, familiarity with possible problems in these systems is essential when troubleshooting a particular instrument malfunction or symptom.

The pneumatic system generally consists of the carrier gas supply, pressure regulator and/or flow controllers, the column and all connecting tubing. The sample inlet system usually consists of a carrier gas preheater and the sample flash-vaporizer. A stream-splitter following the inlet system must be used when capillary columns are employed, as the small inner diameter capillary columns, usually 0.01-0.02 inches, can accept only 1/100 of a microliter of sample. Larger volumes generally are used for analysis with packed columns. A block diagram of the pneumatic and sample inlet systems is shown in Figure 8-1.

Following are brief descriptions of how the pneumatic and sample inlet systems function and suggestions for routine maintenance of the various components.

The carrier gas is usually supplied from a cylinder (Figure 8-1, component A) with the initial pressure of 1800-2400 psig. The cylinder is fitted with a double-stage presure regulator that allows pressure regulation of 10-250 psig

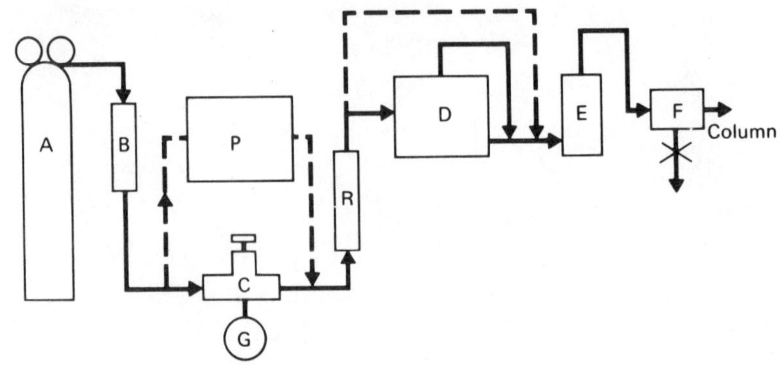

A = Carrier gas supply (usually helium or nitrogen), with pressure reducing regulator.
B = Carrier gas filters.
C = Pressure regulator.
D = Flow controller (with flow control needle valve).
E = Sample inlet (vaporizer).
F = Sample splitter (if capillary columns are used.
V = Split control needle valve.
R = Rotometer (optional).
P = Pressure programmer (optional).
G = Pressure gauge.

Fig. 8-1. A block diagram of the pneumatic and sample inlet system as far as the separating column. (1)

at the 2nd stage outlet. The carrier gases generally used are helium, nitrogen, hydrogen, or argon. Carrier gases are available commercially in various grades of purity. Research-grade or "GC-grade" gases are preferred because they have been refined to contain a very low level of contaminants. Contaminants commonly found in such gases are water, hydrocarbons, CO_2 and other inert gases. Although nominal concentrations of impurities in the carrier gas do not appreciably affect GC retention behavior, the effect of impurities on detector stability and response can be quite serious. The magnitude of this problem increases in direct proportion to the sensitivity of the detector. Thus, it is of utmost importance to pass the carrier gas through a filter or adsorption trap (Figure 8-1, component B) to remove impurities. Carrier gas filters or adsorption traps are usually 6-12 inches long with 3/8 inch or larger stainless steel tubing and are filled with type 5A molecular sieve to remove hydrocarbon impurities and moisture from the carrier gas. When using a good grade of carrier gas, a molecular sieve trap should be effective for at least one year without regeneration or replacement. However, if necessary, the sieve may be regenerated by removing from the tubing and heating to 300°C for eight hours (preferably in a vacuum oven). Charcoal filters are sometimes used for removal of light hydrocarbons and activated silica gel traps are used for moisture removal from the carrier gas.

When changing carrier gas cylinders, it is important to ensure that all fittings are free of dust and dirt particles before assembly, as these materials could enter the gas stream and cause plugging of the small orifices in the flow controllers, valves, etc. Since most gas cylinders are stored out of doors, be especially careful that all rust particles have been brushed away from the valve area before attaching the double-stage pressure regulator. Periodic checks for leaks should be performed using a soap solution. This should always be done when gas cylinders are changed. The major symptom of a leak in this area is the abnormally rapid use of carrier gas. Do not wait too long to change carrier gas cylinders. The outlet pressure of the tank should never be allowed to drop below 100 psig, since running the tanks near empty increases the probability of introducing the impurities (especially water) present in the carrier gas into the GC flow system. This could result in possible disarming of the gas filters and cause intermittent, spurious spikes to show up on the chromatographic recorder as the impurities pass through the detector.

A good rule to follow with molecular sieve traps used for removal of water and hydrocarbon impurities is to regenerate or replace the sieve in the trap twice a year, especially if commercial-grade gases are being used for the carrier gas. When the molecular sieve traps become saturated, impurities in the carrier gas begin to leak through into the gas-flow system and usually cause intermittent spikes to appear on the GC recorder. This problem often appears to be a disturbance in the detector/electrometer system. However, if the spikes are quite random in size and time of appearance, impurities in the carrier gas should be suspected as a possible cause. Replacement of the molecular sieve trap will eliminate the problem. Another effect of water entering the gas-flow system could be the rapid deterioration of certain types of column substrates, such as polyesters, especially at elevated temperatures. Thus, it is of utmost importance to use good grades of carrier gas, and to scrub impurities from the carrier gases adequately to ensure that a pure, dry gas enters the chromatograph.

Pressure regulators (Figure 8-1, component C) are used in some commercial units and in most homemade units. Such regulators allow an accurate pressure supply to the unit, especially when more than one GC is connected to a single pressure cylinder, which could result in differential pressure changes during adjustments at other GC units. These are usually regulators of the nonbleeding, spring-loaded, diaphragm type as shown in Figure 8-2. A pressure gauge is usually an integral part of most pressure regulators.

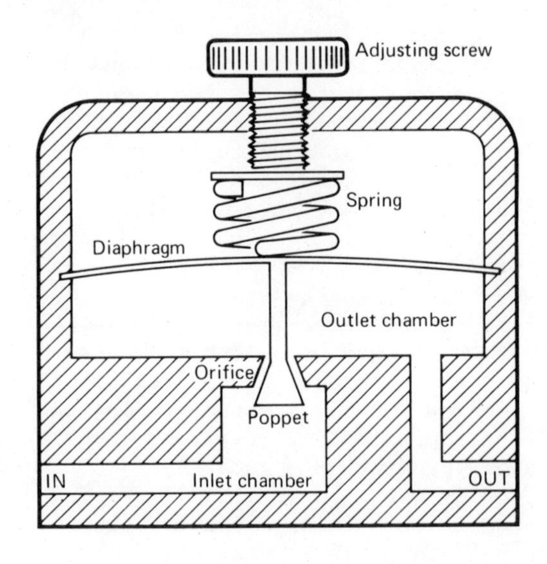

Fig. 8-2. Pressure Regulator (Courtesy of Veriflow Corp.)

The tube and pipe connections in and around the regulator and pressure gauge area should be checked periodically for leaks with soap solution, especially after any kind of tear down and reassembly operation. As in the case of leaks in the pressure cylinder or gas filters, leaks around the pressure regulator and gauge will not greatly affect the GC unit's performance, but can result in the loss of exorbitant amounts of expensive carrier gases. The only part of a nonbleeding pressure regulator that can deteriorate in service and require replacement is the rubber diaphragm. A good check on the condition of the diaphragm is to disconnect the flow-transport line from the inlet of the rotameter (Figure 8-1) and close it off with a plug. Adjust the regulator until about 30 psig shows on the gauge then back-off the adjusting screw (Figure 8-2) until no pressure can be felt on the spring-loaded diaphragm. If the gauge pressure remains constant at about 30 psig, the rubber diaphragm and the pressure regulator-gauge system is leak-tight. However, if the pressure on the gauge drops from the 30 psig setting, a leak is present. If the fittings in the system all test leak-tight with soap solution, a ruptured diaphragm is indicated. The diaphragm is easily replaced by removing the hold-down bolts along the regulator flange, removing the old diaphragm and replacing it with a new one. After this operation, the above leak test should be repeated.

Rotameters (Figure 8-1, component R) are available as inexpensive options for most commercial units, and are used for measuring mass gas flows (calibrated in cm^3/min at various inlet pressures). However, a rotameter is only a convenience to re-establish a given flow rate, as flows are more accurately measured with a simple soap film meter. The main malfunction that could occur in a rotameter would be a leak in the tube connections or the possible accumulation of dirt or excessive moisture on the inner walls. If the carrier gas and components on the pressure cylinder side of the rotameter are kept clean, no maintenance should be required. However, if some foreign material should get into a rotameter unit, it can be disassembled and the calibrated glass tube cleaned with a solvent and dried thoroughly. The floats should be cleaned with soap and water and dried thoroughly. The unit can then be reassembled, recalibrated using a soap film meter, and put back into service.

Differential flow controllers (with flow control needle valve) are present in most modern GC units (Figure 8-1, component D). A schematic of a typical flow controller is shown in Figure 8-3. Once the desired operating pressure has been set at the pressure regulator, the needle valve of the flow controller is used to set the required flow-rate. Once the initial pressure and flow are established, the flow through

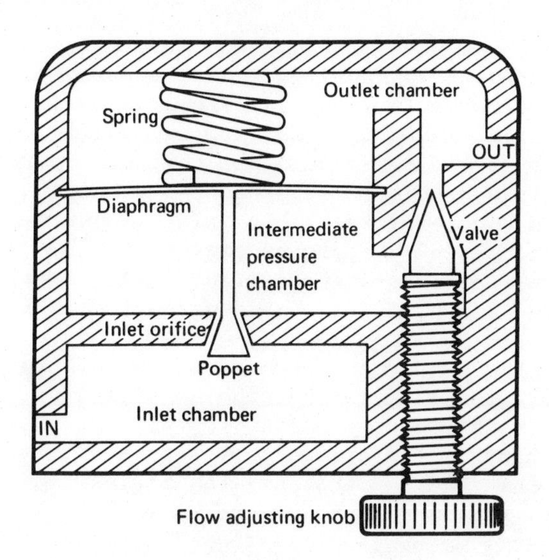

Fig. 8-3. Flow Controller (Courtesy Veriflow Corp.)

the column is kept constant during changing pressure drops (ΔP) across the column, as are caused by the programming the column temperature (the resistance to flow in the column increases with increasing column temperature causing a greater ΔP across the column). The spring-loaded diaphragm (Figure 8-3) of the controller is positioned by the inlet pressure of the incoming gas on one side and is maintained in a given position by a balance with a combination of the spring and the pressure at the outlet. If this balance is disrupted by a ΔP beyond the controller outlet (across the column), the diaphragm will automatically reposition itself, causing the controller valve to open or close, thus ensuring a constant mass flow through the instrument. The end effect is a constant volumetric flow of carrier gas as it leaves the column at room temperature and pressure.

Pneumatic flow controllers such as the one just described have little that can go wrong except for trouble in the diaphragm area or in the controller needle valve. However, controllers will operate well for long periods of time if preventive maintenance is periodically performed. Most flow-controller needle valves are manufactured to close tolerances, and precautions should be taken in their care and use. Never turn the needle valve off too vigorously, this could damage the O-ring shut-off seal and the precisely machined needle. Also, when working on the carrier gas plumbing on the inlet side of the controller, take care to prevent dirt, metal chips or other foreign objects from entering the lines, as these could clog narrow apertures in the controller.

If the inlet pressure to the controller is too low, it will not compensate properly for ΔP changes beyond the outlet of the controller. At least 10 psig is needed between the controller inlet and outlet for proper sensing of small pressure changes across the column for reliable operation. Most controllers can handle inlet pressures as high as 150-200 psig, and as a general rule, the inlet pressure should always be set as high as 60-70 psig to prevent the low-pressure differential malfunctions just described. As usual, the entire system around a controller should be periodically checked for leaks with soap solution.

The best operational check of a flow controller is to set up the unit for a temperature-programmed analysis, and measure the total volume flow at the tail of the column or detector outlet as the column temperature increases. If the flow (cm^3/min) remains constant with increasing column temperature, it may be assumed that the controller is operating efficiently. However, if flow decreases with increasing column temperature, the controller should be disassembled, all lines and orifices thoroughly cleaned the diaphragm checked for pin-hole leaks, and any defective or apparently worn parts replaced. After reassembly, the same test should be repeated to check the performance of the overhauled controller.

The discussion so far has been concerned with the single chromatographic system only. While this is the type of system many people use, there is often an economical and a safety advantage in combining the gas supplies for a number of chromatographs into one gas system. Instead of one gas cylinder for each chromatograph, four cylinders attached to a Matheson manifold system are used to supply 10 chromatographs. This enables one bank of the manifold to be isolated and the two cylinders changed without reducing the line pressure because the other cylinder bank is supplying the required gas flow. Thus, none of the chromatographs need to be shut down in order to change gas cylinders. In addition, the manifold and cylinders should be placed outside of the laboratory in a service corridor or even outside of the building to allow rapid and easy cylinder changes via easy access of cylinders to the manifold. Such a manifold system, in which the chromatographs can be operated during gas cylinder changes, is a great help in obtaining the reliable operation of the GC-units, since no flow shutdown is necessary during gas changes.

There are three pressure gauges installed in the manifold; one regulator for each two-cylinder bank and one master regulator-gauge assembly that is used to set the desired line delivery pressure. A filter system is installed just inside the GC laboratory, and as is the case with the cylinders, duplicate filters are installed in parallel so that only one is on-line at any time. This allows for rapid changeover of filters if contamination is suspected. Such manifold and filter systems are installed for all gases required for GC work in the laboratory, and include helium, nitrogen, hydrogen and analytical-grade air. These gases are piped via 1/4 inch O.D. copper tubing to the various GC-units in the laboratory, with all four gases available at each GC outlet. Each of the gases at each of the outlets is fitted with a single-stage regulator that allows each chromatograph to be operated independently of the line pressure and the pressure requirements of the other instruments. This is not required for chromatographs with built-in pressure controllers, but is deisrable for units which are fitted with flow controllers only.

Leak testing of such a system obviously can be somewhat difficult; however, by following a logical sequence of testing, most leaks can be found quickly, isolated, and repaired, especially with a portable thermal-conductivity-based leak detector that is now available from the Gow-Mac Instrument Company. After changing a gas cylinder bank, that bank of the manifold should be leak-tested each time. Otherwise, it is only possible for leaks to develop at the chromatograph outlets, as solid lengths of 1/4-inch O.D. copper tubing are used between the filters and the various chromatograph outlets. From a safety standpoint, it is recommended that the hydrogen system be checked for leaks every week, as it is unlikely that any leak could develop into a major leak before it is noticed or detected. A good procedure to follow is to pressurize the whole system to the line pressure, isolate the cylinders at the manifold and the chromatographs at the outlets, and record the pressure on the main line pressure gauge at the manifold. If this pressure falls more than one or two pounds per square inch in five minutes, then a sizable leak is apparently present in the system. The rate at which the pressure falls will give some indication of the size of the leak. If the leak is important enough to be tracked down and repaired, and this is obviously a decision made with reference to the particular gas supply being tested, proceed to isolate sections of the line. With hydrogen and oxygen, any noticeable leak should be isolated and checked immediately, whereas small leaks in the helium and nitrogen systems can be tolerated. With the analytical-grade air, somewhat larger leaks are not in any way dangerous, and since the compressed air is reasonably inexpensive, there is not sufficient economic pressure to chase down every small leak detected. If there is a leak which needs repairing, then the following procedure is used: (1) Each leg of the system is isolated in turn from the main supply line which contains the pressure gauge and the line pressure is watched closely. (2) When the system leg that contains the leak is isolated from the gauge, no further falling of the pressure on the main gauge should be noticed. (3) Once the leg containing the leak is determined, then proceed by checking, first of all, any couplings which are occasionally opened on that line, the packing nuts of any valves, and any other couplings in the line. It is very rare that a leak is found which occurs somewhere other than a coupling, and therefore, when assembling a system, the minimum of couplings should be used. (4) Leak repair usually consists of tightening up the coupling, or, in extreme cases, in cutting off the coupling nut and ferrules, and replacing it with a new one. The hardware used in this manifold system (regulators, controllers, and filters), are exactly as described earlier in this chapter.

Leak testing chromatographs is usually confined to the times when the column is changed or if the flow characteristics through the chromatograph change. This is easily done with the helium leak detector without disturbing the operating temperatures of the inlet system or the detector. If no leak detector is available, again pressure test the system by blanking off the end of the column (the end which is usually connected to the detector) and pressurizing the system up to the line pressure and recording the pressure on the column head gauge. This is done with the column and the injection block at operating temperatures. As in many cases, leaks on the GC unit itself are not noticeable when operating at room temperature, but become quite appreciable at temperatures normally used for analyses. Leaks through the septum are easily found with the leak detector, or by a technique that will be discussed later in this chapter when septa are the sampling-system component under scrutiny.

It is the function of a well designed sample inlet system (Figure 8-1, Component E) to receive the sample, vaporize it instantaneously, and deliver the vaporized material to the head of the analytical column in a narrow "plug," as elution band (or peak) widths are directly related to the injection-band width (2). Thus, effective resolution of close boiling materials could be lost by an inferior injection system. A good example of a flash vaporizer with a removable liner is shown in Figure 8-4 (3).

As in all well-designed vaporizers, the carrier gas is preheated by passage through a short length of tubing or a bored-out part of the vaporizer block before coming in contact with the sample. The added preheater volume also acts as a flow-surge buffer for the inlet system. To ensure "plug" flow to the head of the column, the carrier gas enters the inner liner at the front so that the sample will be swept toward the head of the column, minimizing any chances for back-diffusion or band-broadening in the vaporizer itself.

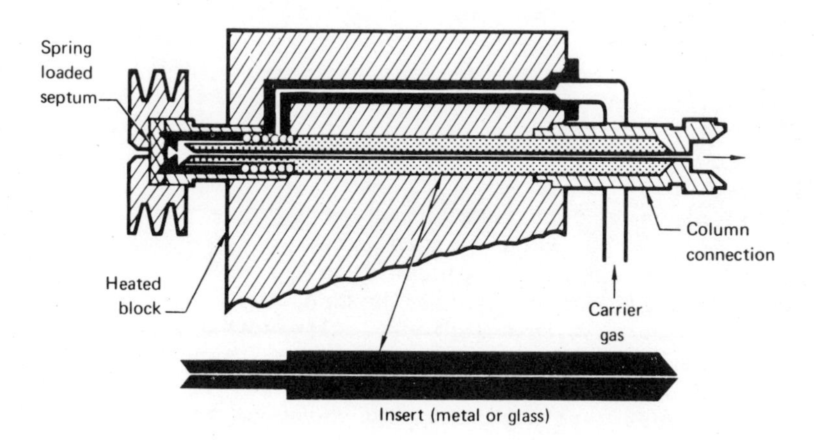

Fig. 8-4. Typical liquid sample inlet system (Courtesy of Perkin Elmer Corp.) (3)

This type of vaporizer is often referred to as the concentric-tube configuration. The inner liners are designed with a small internal volume (2-4 inches long by 0.028-0.030 inches i.d.) to help ensure "plug" introduction of sample at the head of the column by keeping vaporizer residence time to a minimum. However, such a high-temperature stainless-steel vaporizer can only be used with thermally stable samples that do not react with the hot metal surfaces. The same type of inlet system can be used with some unstable materials by substituting a glass liner for the stainless steel liner. Several commercial chromatographs have sample inlet system vaporizer designs similar to that shown in Figure 8-4.

The most common method for introducing liquid samples into the chromatograph is by a microsyringe. The desired sample sizes are usually between 1 and 10 microliters for analytical work, thus 5 or 10 microliter syringes are generally the most popular. The sample is introduced with the syringe through a silicone-rubber septum into the vaporizer. The vaporizer is normally maintained at 250 to 350°C. At these temperatures, silicone septa begin to degrade rapidly and must be changed quite often to prevent leakage. High-temperature septa that are more resistant to thermal decomposition are now available. Also available are laminar constructed septa, with the layer in contact with the inside of the vaporizer made from polytetrafluoroethylene (PTFE) to decrease possible adsorption of components by the septum.

At this point, a few words concerning the probable pitfalls for the chromatographer using syringe injection through rubber septa are in order. In this era of GC unit development, when chromatographers as a group are looking for methods to accurately determine lower and lower (trace) amounts of various components, the contamination role played by standard, off-the-shelf septa must be considered, especially when large amounts of solvent are used to inject very small amounts of heavy or insoluble materials. As a case in point, consider the following problem with septa encountered in an industrial quality-control laboratory.

Oil mists in industrial atmospheres were routinely sampled in charcoal tubes and returned to the laboratory for analysis. The oil was displaced from the charcoal with a known volume of solvent (either hexane or toluene). This solution containing the displaced oil was then analyzed by GC, and the integrated record compared to standard samples of oil base stocks in toluene for quantitation. The septa used in the analyses were high-temperature septa from a well-known GC supply house. The inlet temperature was maintained at 300°C. Analyses of the first samples of displaced oil mists were uneventful, with the hydrocarbon profile of the oil very similar to that obtained for the base stock standards. However, strange peaks began to appear on the top of the hydrocarbon envelope normally observed for the oil base stock, and there was much speculation as to what kinds of components had been trapped from the industrial air. About five more air samples were analyzed, and the same large and quite high-boiling peaks appeared again and again. The next morning, a blank run (temperature programmed with no sample injected) was made, and the large peaks, minus the oil base stock profile, appeared again. Contamination of the septum was immediately suspected. So, a piece of aluminum foil was placed behind the septum, and the blank run repeated. No big peaks were observed! One of the air samples analyzed the first day was run again, and no large peaks were observed, only the peaks that could be attributed to the oil base stock were found. Thus, the large amount (10-μl) of solvent in contact with the septum at 300°C was leaching-out heavy materials from the septum body exposed to the inlet system, and these materials were stored on the head of the column until the column temperature was programmed to a sufficient level to start them moving through the column with the oil base stock or without the presence of the oil base stock. With this clue, several of the same septa were extracted with hot chloroform on a steam plate, and then stored in a vacuum oven at 100°C overnight, and thereafter. From this time on, the septa were handled only with clean tweezers when being placed into the septum holder of the GC inlet system. The same oil mist samples were analyzed again, and the ghost peaks were no longer in evidence, even on a current range of 10^{-11} amperes for the electrometer monitoring the output of the flame detector. Thus, it is recommended that the solvent extraction -- vacuum storage -- transfer treatment be followed for all septa to be used in trace analyses at inlet vaporizer temperatures of 200°C or higher.

Malfunctions in the sample vaporizer can completely impede any reliable GC analysis. Routine or periodic maintenance of this system is a must; therefore, some of the commonly occurring malfunctions will be discussed briefly. If the inlet system is dirty, the interior walls of the system that come in contact with the vaporized sample can become covered with a thin layer of heavy coke-like material. This is especially true for samples such as gasoline that contain extremely heavy and nonvolatile detergent additives that are not vaporized at normal inlet temperatures. After a while, this material can become hardened and coke-like, as any slightly volatile materials are slowly swept away by the carrier gas. Such a dirty inlet can cause several problems which are listed below.

1. Peak broadening or shifting resulting from

interaction of components with active (polar) sites on the deposit or caused by adsorption effects from the porous nature of the deposits.

2. Sometimes complete removal of polar or unstable material will occur because of interaction with the deposit at temperatures above 200°C.
3. Ghost peaks (peaks not related to the sample being analyzed) will appear on the recorder. These could be materials being desorbed from the inner lining of the vaporizer. This problem can usually be solved by retubing or replacing the inner liner of the vaporizer.

A leak in the inlet system can be one of the most important system malfunctions to the worker who is trying to obtain consistent high-resolution quantitative analyses, mainly because inlet system leaks are hard to find due to the extremely high temperature of the system. If the inlet system is cooled to room temperature, the leak may disappear as the metal blocks and tube fittings contract on cooling, possibly stopping the leak. Thus, the most expedient way to check an inlet system for leaks is to block off the exit to the column with a tubing plug, pressure-up the system to about 50 psig (or to the maximum pressure at which operation will be carried-out), and close off the pressure regulator completely (as described in leak checking the pressure regulator/gauge system previously in this chapter). However, this technique is useful only if the pneumatic system preceding the inlet system is leak free. If no loss of pressure is observed the inlet system is not leaking, and the trouble should be looked for elsewhere in the chromatographic system. Be sure to install a new septum before conducting this test, as a leaking septum could appear as a leak in any other portion of the inlet system. Leaks can result in the following overall symptoms: peak broadening, loss of resolution, loss of more volatile materials relative to heavier ones, and the appearance of ghost peaks.

A leaky septum can produce all of the symptoms listed above. The best method of routine maintenance and operational checks that can be suggested here is to change the septum frequently. Septa are cheap compared with the cost of an individual chromatographic analysis, and a good rule to follow is to change the septum after every 10 analyses when in routine use, or weekly if used only periodically, but with the inlet system temperature maintained at the operating level. So to be safe, change the septum often. A quick and simple way to check for a leaky septum without having to cool down the inlet system is to fit a piece of 1/4 or 1/8 inch o.d. PTFE tubing to a soap-film meter. The septum can then be checked for leaks by tightly pressing the open end of the tube against the septum, and checking for flow with the soap-film meter in the usual way. Another way to detect a leaky septum is the telltale noise of gas escaping when withdrawing the syringe from the septum.

Another malfunction that is often apparent in inlet system vaporizers is the problem of cold spots (i.e., a very small area that is not heated to the same extent as the surrounding vaporizer unit). Cold spots in the inlet system are especially deleterious in the analysis of extremely high-boiling materials. The magnitude of this problem is best shown with an example.

> The GC analysis of 2,4-dinitrophenylhydrazone (2,4-DNPH) derivatives of C_1-C_8 aldehydes was attempted using a chromatograph fitted with an old design inlet system that had about 10-inches of 1/8 inch o.d. stainless steel tubing following the initial vaporizer to provide an adequate mixing volume for the carrier gas and sample. The complete inlet unit was heated to a tubing skin temperature of 350°C. Various columns and many sets of operating conditions were tried to analyze a standard mixture of the C_1-C_8 2,4-DNPH derivatives in benzene solution; however, the only derivative detected was that for formaldehyde (the lowest melting). On reappraisal of the entire system, it was postulated that the sample was probably being lost to a cold spot somewhere in the inlet system. Thus, to remedy the situation, the long, coiled inlet was replaced by a short glass-lined injection port heated to 350°C. This was connected directly to the head of the column. With this improved inlet system, 2,4-DNPH derivatives as heavy as that for tolualdehyde were detected with no difficulty. The failure of a similar old design inlet system may have led Fedeli and Cirimele (4) to draw the conclusion that GC separation of 2,4-DNPH derivatives was probably impossible.

If the samples to be analyzed are somewhat thermally unstable, another method that should be considered for placing a plug of sample at the head of the analytical column is the technique of "on-column" injection, where the sample is placed by various techniques in actual contact with the column packing or coating without the aid or dead-volume of a flash vaporizer. Such a technique is invaluable for GC analyses of materials that would not remain intact when subjected to the thermal shock of the vaporizer. In the technique usually employed, the inlet vaporizer is modified so that the head of the column can be placed about 1/8-inch from the septum. Thus, the syringe needle extends at least two inches into the column for "on-column" injection. One word of caution if this type of injection is to be used. *Do not heat the injection block above the temperature limit of the liquid phase in the column (whether packed or capillary).* If the temperature does exceed the limit of the stationary phase, some phase will be stripped from the column and result in recorder baseline drift or large, skewed peaks.

A method used for "on-column" injection into preparative size columns is through the inlet system to the head of the column via a six-inch syringe. This technique is generally useful for larger sample sizes in the 20-500 microliter range used in analytical/preparative work.

A modified injector insert has been described by Willis and Engelbrecht (5) for "on-column" injection into larger diameter open-tubular columns. This technique is especially useful for analyses of wide-boiling, temperature-sensitive mixtures. Sample sizes used are generally in the range of 0.1 μl or smaller.

A possible bad effect on resolution and general column performance when using any of the on-column injection techniques is excessive peak tailing which may result from adsorption of the sample on uncoated solid support or column walls. This especially becomes a problem after the stationary phase at the head of the column becomes somewhat eroded, and this can be caused by constant disturbance of the packing with the syringe needle.

Other types of sample inlet systems that must be considered are those used for sampling gaseous materials. In areas where the gases to be sampled are under low pressures, a gas-tight syringe of the desired size (usually 0.1 - 5.0 ml) is employed with the usual septum-type inlet system.

If gas samples are to be taken repeatedly in areas from which quantitative data must be derived, a gas sampling valve, fitted with the desired size of sampling loop, is often used. Several commercial gas sampling valves are available, and typical configurations of these valves are shown in Figure 8-5.

Some operate on the slider with O-ring principle (Figure 8-5A), while others, such as the Carle valves, operate by rotation of various flow paths (Figure 8-5B). The major goal when using gas sampling valves is generally good sample size repeatability. This is well controlled by filling the sampling loop at the same pressure for each analysis. In addition, most gas sampling valves may be heated to about 200°C, eliminating any chance for condensation of traces of heavier materials in the sampling loop or connecting tubing. This accounts for better repeatability.

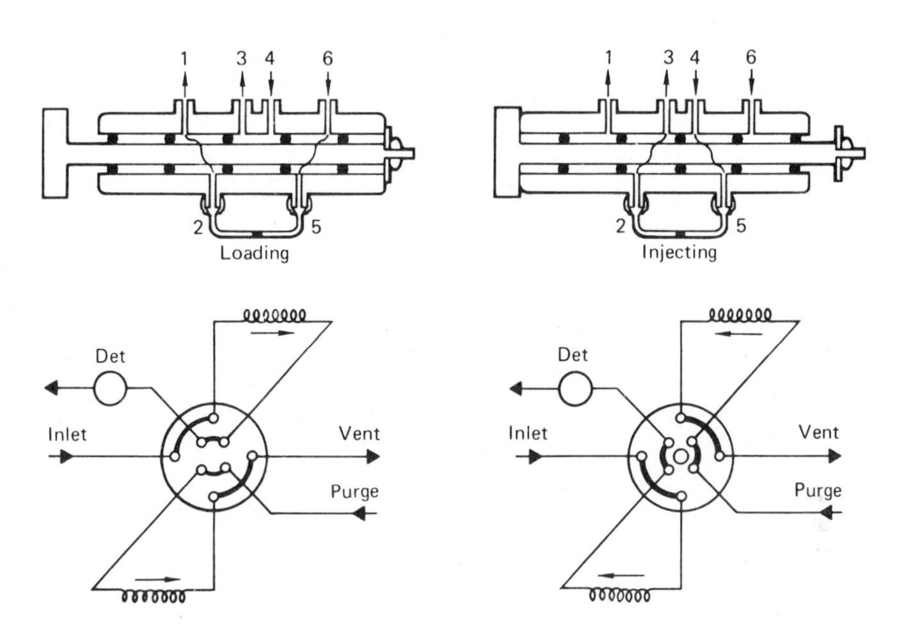

Fig. 8-5. Flow schemes of various gas sampling valves (Courtesy of Varian Aerograph).

An important item in gas sampling valve use is leak-free operation. This is usually obtained by frequent replacement of O-rings in the slider-type valves, and good clean-outs at 200°C+ for the Carle-type valves. Cleanliness of the sampling loop is another important factor. This can be maintained by occasional removal and solvent flushout, followed by thorough drying of the loop before reinstallation.

Another method of gas sampling that should be considered is the technique of encapsulating a gas sample in a small length of indium tubing (6). The indium rapidly melts at a fairly low temperature (157°C), and will instantaneously release an encapsulated gas sample. The gas in question may be sampled by flowing it through a short length of indium tubing. Then, using a pair of pliers, the exit and inlet of the tube are crimped off, resulting in a leak-tight seal, with the gas to be analyzed securely trapped inside the indium tubing. This technique can eliminate, in many cases, the use of steel sample bombs which sometimes rupture or leak because of inadequate venting. The indium capsule filled with the gas to be analyzed is then introduced into the heated inlet of a chromatograph via a grooved plunger assembly with a high-temperature O-ring seal as shown in Figure 8-6. Once in the heated inlet, the indium rapidly melts, releasing the gas sample into the GC sample flow system. Although this

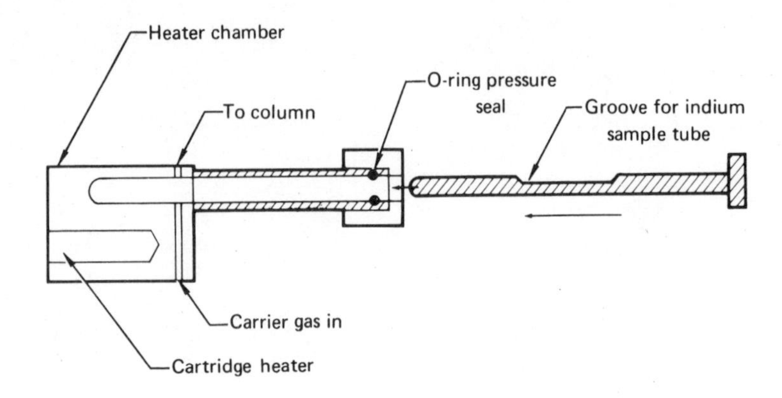

Fig. 8-6. Indium tubing sampling system (Courtesy of Hydrocarbon Processing) (6).

technique makes sampling a relatively easy job, it is difficult to repeat sample sizes, and all work should be done on a relative %w or %v basis.

Referring again to Figure 8-1 (Component F), a sample splitter is necessary if 0.01 or 0.02-inch i.d. capillary columns are used, as capillary columns will efficiently separate only about 1/100 or less of normal sample charges for packed columns, resulting in only about 10 micrograms of total sample entering the capillary column. A common design for a dynamic stream splitter is shown in Figure 8-7.

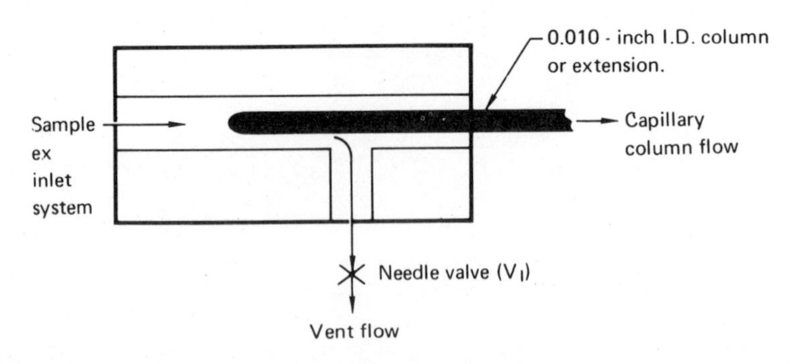

Fig. 8-7. Typical dynamic stream splitter (1).

The inlet of the capillary column (or more generally, a short piece of uncoated 0.01-inch i.d. capillary tubing) is placed in the flowing gas stream ahead of the splitter vent port as shown in Figure 8-7. The ratio of sample entering the column to that being exhausted through the vent is proportional to the flow-rate ratio, or split ratio.

$$\text{Outer Split Ratio} = \frac{\text{Vent Flow}}{\text{Column Flow}}$$

The split flow is readily adjusted by a good stainless steel needle valve (V1 in Figure 8-7). To prevent any prefractionation or condensation of heavier materials in the sample relative to lighter components, the splitter and transfer lines must be maintained at about 300°C. Such nonselective behavior is often referred to as the "linearity" of the sample splitter (7, 8, 9).

A complete checkout of the stream-splitter is difficult because of the high temperature at which it must be operated. The best way to quick check a splitter is to determine its linearity for a wide-boiling standard mixture, employing different sample sizes. If, previously, a splitter had been determined linear for a given mixture, and then, under the same operating conditions, behaves in a nonlinear fashion, there is a leak or deposit problem in or around the splitter and steps should be taken to clean and retube the splitter system. Since the entire system should be made of the same type of stainless steel, a leak-tight system at room temperature should remain leak tight at 300°C as all components have the same coefficient of thermal expansion.

To protect a column that contains stationary phase material with fairly low temperature limits, the column head-pressure can be programmed to speed the elution of heavier compounds once the temperature limit of the stationary phase has been reached (10, 11, 12). If a pressure (or flow) programmer (Figure 8-1, Component P) is in use, the differential flow controllers must be taken out of the system or they would compensate for any flow increases made by the flow programmer. The pressure programmer should be connected directly to the inlet system as shown by the dashed lines in Figure 8-1. A typical piping diagram is shown in Figure 8-8.

Referring to Figure 8-8, the initial column flow is set by the pressure regulator with the start and vent valves closed. The pressure (flow) program rate is controlled by adjustment of the needle valve. To begin a program, the start valve is opened. The carrier gas leaking through the needle valve builds up a pressure in the buffer vessel that actuates a valve in the flow controller (as discussed previously in this chapter), and increases the column head pressure. The buffer vessel combined with the accurate needle valve constitutes a pneumatic time constant (13).

Periodic maintenance of a flow programmer is mainly recalibration since they are entirely pneumatic devices.

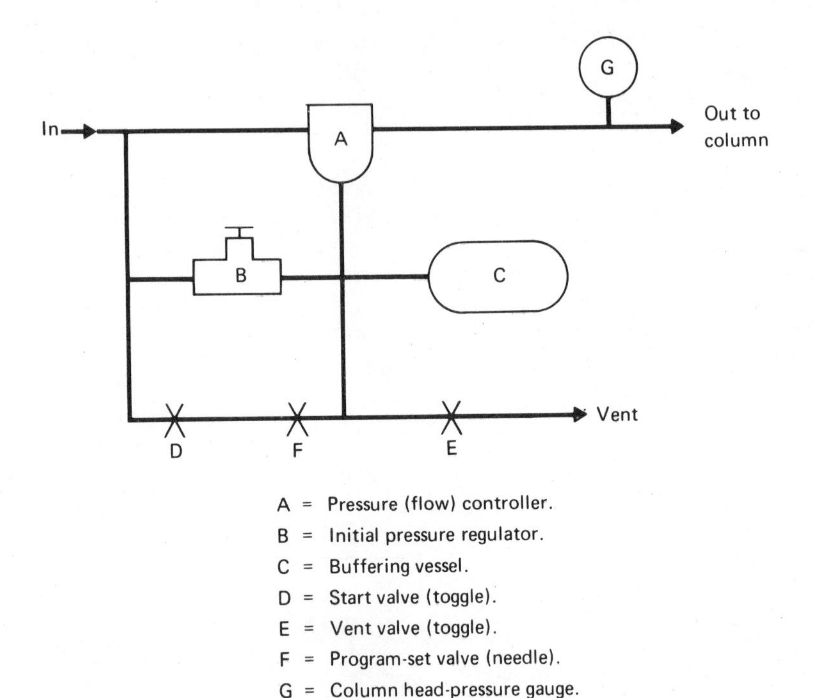

Fig. 8-8. Pressure programmer flow scheme (Courtesy of Analabs, North Haven, Conn.) (13).

However, if dirty carrier gases are used, moisture and/or particulate matter could clog small orifices, causing erratic behavior (i.e., deviation from previously calibrated flow programs). The same principles of cleanliness and periodic leak checks described earlier in this chapter should also be used with the flow programmers. A word of caution concerning the use of flow-programmers; determine whether the detector being used is sensitive to the flow-range being employed (14). If so, nonlinear results will be obtained unless added carrier gas is used after the column in order to make the column flow change small compared with the total gas flow through the detector.

An example illustrating the effect of flow programming on the time required to separate a mixture is shown in Figure 8-9. The mixture consists of C_9 to C_{14} unsaturated hydrocarbons with small amounts of saturated hydrocarbons. The isothermal (125°C), constant flow (12 cm^3/min.) separation of the hydrocarbons required 14 minutes (see Figure 8-9A). Note that the trace saturated hydrocarbons are barely seen. Figure 8-9B is a chromatogram of the same mixture (flow rate 12 cm^3/min), separated at a temperature programmed rate of 10°C/min. The analysis time is 7 minutes, and the alkanes are

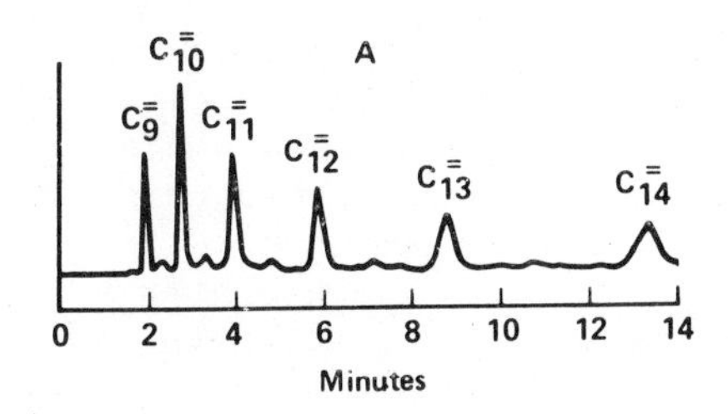

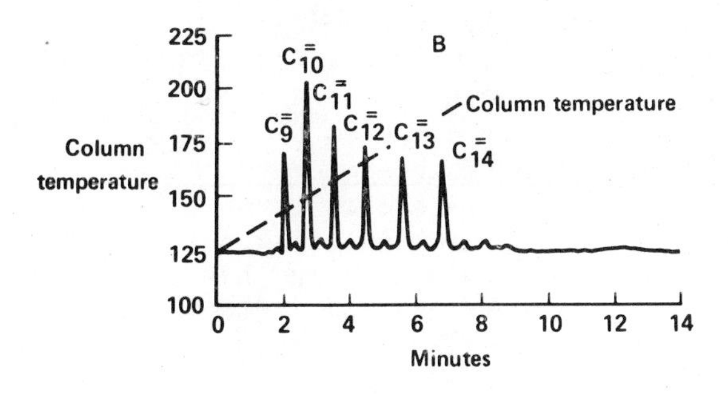

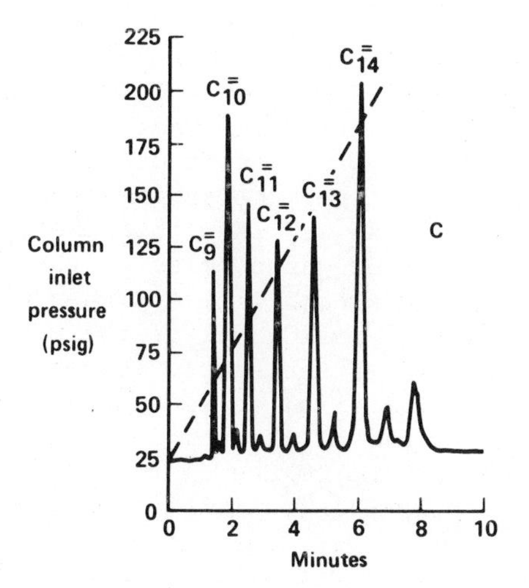

Fig. 8-9. Chromatograms for the separation of a mixture of unsaturated hydrocarbons. (A) Column operated isothermally at 125°C with a constant flow of 12^3cm /min, (b) Column temperature programmed at a rate of 10°C/min with a constant flow rate of 12 cm^3/min, and (C) Column operated isothermally at 125°C while flow programming at a rate of 0.042 atm/sec. (Courtesy of Amer. Lab.)

somewhat more pronounced in Figure 8-9B than in 8-9A. Figure 8-9C illustrates the flow programmed separation (rate of increase: 0.042 atm/sec) of the mixture held isothermally at

125°C. Again, the separation requires 7 minutes, however, the saturated peaks are even sharper than with temperature programming. Moreoever, the areas of the higher boiling, i.e., C_{13} = and C_{14} =, are much larger than those shown in Figure 8-9A and 8-9B. This phenomenon is a direct result of the forementioned change in flame ionization detector sensitivity as the ratio of hydrogen to carrier gas (helium), changes during the analysis. The change in sensitivity may or may not be an advantage to the analyst, depending upon the particular analysis being conducted. This change in sensitivity associated with flow programming can be circumvented by using a carrier gas mixture of 60% N_2 and 40% H_2 (14).

Since the most common system used in gas chromatography for sample introduction is the microsyringe, some cautions as to syringe care and use are in order. The syringe is used for both liquid and gas samples. Probably greater than 80% of all liquid injections are made using the microsyringe. Gas chromatographic microsyringes fall into three categories: First, there are the syringes composed of a lapped metal plunger sliding within a ground glass barrel. The second type of syringe is where the plunger has a Teflon tip which seals against a smooth glass wall, and the third type is where the entire sample volume is contained within the needle.

Filling a syringe varies somewhat according to the type of syringe being used. With the lapped barrel syringes, it is usually necessary to pump the syringe some to wet the surfaces before a sample can be drawn into the syringe without getting a large amount of gas bubbles in the sample. This is not necessary with the smooth vial syringes which can usually be filled with a single withdrawal of the plunger. Pumping of the syringe type where the sample is contained within the needle is usually detrimental to repeatable sample sizes, leads to hold up of some of the sample within the needle and eventually to choking and coking of the needle.

Two types of sample injection techniques are usually employed when using microsyringes. In a normal injection, both sample and air are contained within the syringe. Another method, the "solvent-flush injection technique," is mainly applicable to samples containing high-boiling components, and particularly should be used for samples where one of the components is a solid dissolved in the liquid solvent. Filling a microsyringe for these types of injection are done as follows: First, for the normal injection, the microliter syringe is dipped into the sample and the plunger pumped up and down two or three times in order to wet the surfaces and remove the majority of air bubbles from the sample. The plunger is then slowly withdrawn and some sample withdrawn into the syringe. It is usually preferable to overfill the syringe at this stage. The syringe is then removed from the sample and turned so that the needle is in a vertical (upward) position. The plunger is slowly depressed until the required amount of sample, as indicated on the syringe scale, is contained within the syringe. The plunger is then withdrawn slightly in order to draw some air into the barrel of the syringe. This serves to remove sample from the needle, and thus, minimizes the chance of pre-volatilization of the sample once the needle is inserted into the chromatograph inlet system. With the "solvent-flush method," the syringe is first of all dipped into the solvent being used and a small amount of the solvent (about 1.0 µl) pulled into the syringe. The syringe is withdrawn to pull in a small 1.0 µl slug of air. Then, the syringe is dipped into the sample, the sample withdrawn, and finally, air is drawn into the syringe as in the first sampling method described. The purpose of this technique is so that when the sample is injected into the chromatograph, the needle is flushed with solvent, and, therefore, any of the high-boiling components which have tended to deposit in the needle are flushed out with the follow-up of solvent. This technique is particularly good in dealing with almost any materials which are solids at room temperature.

Following is a list of things to remember when using syringe injection.

1. Be sure that the syringe is clean. The sample sizes injected with microliter syringes are so small that contamination can be a major problem. Therefore, it is best to insure that the syringe is clean immediately following the sample injection, rather than waiting for the next sample before cleaning the syringe. So, clean the syringe as soon as possible after injecting the sample.
2. Be certain that the syringe is in good condition. This includes making sure that the needle is straight, that it is not blocked, that the plunger moves easily within the barrel, and that it still makes a vapor-tight seal so that the vaporized sample is not going to be blown back past the plunger.
3. Be sure that the syringe is the right size for the job. Do not try to inject very small samples with a large size syringe. If the sample size required is 1.0-µl, then the largest syringe that should be used is a 10-µl syringe, and preferably a 5.0-µl or 2.0 µl syringe should be used when injecting a 1.0-µl sample. The risk of poor sample injection is also great if a syringe is used to its full capacity. For instance, to use

a 5-μl syringe to inject 5-μl's of sample is a risky business. In all cases, syringes perform best if the sample size injected is between 20 and 80% of the total marked volume of the syringe.

4. Be sure to wet the inner surfaces of the syringe by pumping the plunger up and down while drawing in sample. This is particularly needed for the lapped glass barrel syringe. It is not necessarily needed with the other syringe types. Overfilling of the syringe is also valuable in order to remove air bubbles from the sample, and thus, to enhance the repeatability of the injection. Make sure that the outside of the needle is clean and dry. Contamination from the outside of the needle is particularly noticeable when dealing with small diameter columns where an adequate sample size could actually be obtained by dipping the syringe into the sample and injecting, with no need to contain any of the sample within the needle of the syringe. In order to get repeatable sample sizes onto packed columns, it is important to insure that no sample is on the outside of the needle.
5. When injecting the sample, one should be very sure to see that the syringe needle is inserted into the heated injection zone quickly. This minimizes the chance of the sample boiling off from the syringe before the sample has been completely injected, and thus, contaminating the septum and getting a broad band injected onto the column. The technique usually used is to insert the syringe through the septum and then quickly push the syringe-needle all the way into the vaporizer until the barrel of the syringe is resting against the septum or the septum-retainer nut. Only then is the plunger depressed and the syringe needle removed from the injection port immediately after the sample has been injected. This serves to squirt the sample into the heated zone of the injection port and aid in the desired instantaneous vaporization of the sample.

Maintenance of gas chromatographic syringes is a never-ending business, which starts immediately after the injection of a sample. In order to keep the syringe in good condition, it should be flushed carefully with solvent so as to remove any last traces of the previous sample. Then, it should be blown dry by inserting the syringe into a clean air line so that gas blows through the needle and out through the barrel of the syringe. This technique is to be preferred over a system where the syringe is inserted into a vacuum line, because in the latter case, dirt from the rod can be sucked through the barrel of the syringe, leaving a characteristic black ring around the end of the syringe. In cleaning syringes, one must be very careful not to handle the plunger with the hands, as traces of perspiration left on the plunger will corrode the glass barrel and possibly cause the syringe to stick. Other maintenance items to remember with syringes are to develop a smooth technique for injecting the sample, and thus, minimize the problem of bending the needle and blocking it with the septum material, and/or bending the plunger. This usually occurs when the sample is being forced into the chromatograph too quickly, particularly if the injected material is quite viscous. The usual result of trying to increase the speed of injection of a viscous material is to bend the plunger, but in extreme cases, hydraulic pressure built up in the syringe can be so great that the glass body of the syringe can crack.

Properly cleaned and maintained, a microliter syringe can last for several years. Clumsy use of the syringe, however, can usually manage to destroy its usefulness during the first series of injections, and this becomes quite an expensive way to do gas chromatography. In addition, syringes that are stuck or have bent plungers can usually be sent back to the manufacturer for reconditioning at about 1/3 of the cost of a new syringe.

In this chapter we have described the operating principles of the individual components that make up the pneumatic and sample inlet systems of a gas chromatograph, and discussed routine (or preventive) maintenance procedures for each component in these two systems. Since malfunctions in the pneumatic or sample inlet systems can appear as problems in other instrument systems, a thorough understanding of the operation of the various components in these two systems, as well as symptoms they can produce when not operating properly, is essential.

REFERENCES

1. J. B. Maynard, Private communication, 1971.

2. S. DalNorgare and R. S. Juvet. *Gas-Liquid Chromatography - Theory and Practice*, 3rd ed., Interscience Publishers, New York, 1965 (p. 166).

3. Literature on Oven Designs for Gas Chromatography, 1967, Perkin-Elmer Corporation, Norwalk, Conn.

4. Fedeli and Cirimele, *J. Chromatog.*, *15*, 475 (1964).

5. D. E. Willis and R. M. Englebrecht, *J. of Gas Chromatog*, August 1967.

6. J. Q. Walker, *Hydrocarbon Processing*, *46*, No. 4, 122 (1967).

7. L. S. Ettre and W. Averill, *Anal. Chem.*, *33*, 680 (1961).

8. L. S. Ettre, E. W. Cieplinski and N. Brenner, "Quantitative Aspects of Capillary Gas Chromatography," ISA Reprints, No. 78-LA/61 (1961).

9.- L. S. Ettre and F. J. Kabor, *Anal. Chem.*, *34*, 1931 (1962).

10. J. D. Kelley and J. Q. Walker, *J. Chromatogr. Sci.*, *7*, 117 (1969).

11. J. D. Kelley and J. Q. Walker, *Anal. Chem.*, *41*, 1340 (1969).

12. C. J. Wolf and J. Q. Walker, *Amer. Lab.*, *3*, 10 (1971).

13. Literature on Analabs Pressure Programmer, 1969, Analabs, Inc.

14. R. L. Levy, J. Q. Walker, and C. J. Wolf, *Anal. Chem.*, *41*, 1919 (1969).

QUESTIONS AND ANSWERS

Question: At exactly what distance in the hydrogen line should "flow snubbers" be installed?

Answer: Hydrogen "flow snubbers" should be installed just after the pressure relief valve and before the GC unit. In addition, all hydrogen connections (except the connection to the chromatograph itself) should be made outside of the laboratory in a well ventilated air space. *One piece* of copper or stainless steel tubing should be run from outside the building (or service corridor) to the laboratory if possible.

Question: When using the "flush" method for injecting samples, is the amount of air injected with this technique detrimental to some GC stationary phase?

Answer: The air in the sample would be bad for the stationary phase only in large amounts and with constant repetitive injections onto phases that are susceptible to oxidation. In general, such a technique, where about 1.0-µl of air is injected with each sample, is not harmful to most stationary phases.

Question: When a fairly large sample (10-µl) is injected, the rotameter float drops several divisions, then rather quickly returns to the normal level of flow. What can cause this?

Answer: This fluctuation in the flowmeter reading is probably caused by backpressure due to rapid vaporization of the sample in the GC inlet system. This effect can usually be eliminated by reducing the sample size, increasing the carrier gas pressure entering the inlet or reducing the vaporizer temperature, if possible. However, the controlling factor is mainly the volatility and sample size of the material being injected. If the peak shape in the resultant chromatograms are good and the retention times are repeatable, it is best to forget about the backpressuring, as long as it is not detrimental to the analysis.

Question: How useful are backflush valves? Can their use significantly reduce analysis times in wide-boiling materials?

Answer: Backflush valves are very useful, and can cut analysis times considerably, especially if you are only interested in analyzing for the amount of light components in a wide-boiling mixture. For example, only the C_5 components in a refinery stream are of interest; thus, after the elution of the C_5 materials, the rest of the sample can be determined very quickly by backflushing out the heavy ends. Such an analysis scheme is very common in refinery quality control work.

Question: What is the correct technique for using a microsyringe for reproducible sample injection volumes?

Answer: To achieve this objective, flush the syringe up and down in the sample to wet the syringe walls and remove most of the air bubbles. Strive to inject the sample with the same syringe stroke each time, with a smooth needle insert through the septum and a smooth needle withdrawal. Too rapidly withdrawing the needle can allow some of the vaporizing sample to escape through the septum as the syringe is being removed. Some workers prefer the "flush" method of injection as previously described in this chapter. However, both techniques can be used to good advantage, depending on the skill of the chromatographer.

Question: What size of tubing in a gas supply system would be necessary to feed at least three chromatographs?

Answer: A 1/4-inch O.D. tube is more than adequate for supplying three GC units with carrier gas. In our experience, 1/4-inch O.D. copper tubing is used to run from a cylinder manifold to 10-12 GC units with more than enough capacity at an inlet pressure of 60 psig. In this system, the manifold is placed at least 50-feet from the actual GC laboratory.

Question: Why would a sample hang-up one time in a column and not hang-up during the next analysis?

Answer: This could be indicative of a poor septum or poor injection technique. Sample hang-up could mean sample-loss by backflushing through the septum, especially if the problem is intermittent. This could also point to quite inconsistent injection techniques. What appears as sample hang-up (or absence of sample at the detector) occurs over a given period of time. The simplest way to check out such a situation is to run a standard material that is easily analyzed under the conditions being used. If the standard material appears as expected, then there is a possibility that sample hang-up is actually sample decomposition in the inlet system or on the column, and the severity of this problem may vary from time to time with the cleanliness of the inlet system and the head of the GC column.

Question: A large difference in retention times and peak shapes was noticed when using a 500-foot x 0.02-inch I.D. wall-coated open-tubular column in two different commercial GC units. All measurable operational variables were the same in both cases. No splitter was used with a 0.1-μl sample size. What could cause this behavior?

Answer: Either the unit giving the poor peak shape has a leak in the inlet system or the inlet system of the unit giving poor peak shapes is of poor design for use with a 0.02-inch I.D. column, probably with far too much dead volume for use with a column of this size. Probably, the dead-volume possibility is most likely, as both instruments were carefully checked-out by company service engineers. An injection port with too large an I.D. or a possible area unswept by the carrier gas could definitely result in the symptoms described. Also, there is the possibility of uneven heating of the column during the temperature programming in the unit, giving the poor peak shapes.

Question: Do smaller dimensional components such as valves, connectors and columns generally decrease retention times?

Answer: Yes, smaller dimensional components decrease instrument dead-volume, thereby, generally decreasing retention times and overall analysis times if mated to, say, a desirable SCOT column. Many older process units can be converted to more rapid operation at the same or better resolution by decreasing sample sizes and all the internal volumes of the GC flow system.

Question: Two identical 1/8-inch O.D. columns were prepared from the same packing material for a specific analysis of cyclohexane in a light chlorinated solvent. One of the columns gave a good peak for the cyclohexane while the other gave a sharp, doublet-type peak. What could be causing this difference in behavior between the two columns?

Answer: If this occurrence was observed only once or twice,

the sample injection technique should be suspected, as "doubling" sometimes occurs for fairly small amounts of materials in a large excess of solvent, especially if some trouble is encountered in cleanly injecting the sample through the septum. However, if this occurs repeatedly, the column rather than the sample injection, should be suspect. There could possibly be a leak at the head of the column, or channeling throughout the column producing inconsistent resolution; however, this is much less likely than poor sample injection.

Question: What is the best way to get repeatable injections of samples to get good quantitative results from GC data?

Answer: Repeatable injections can be obtained by using the same technique very carefully and consistently. If analyses are based on an internal standard, a fairly consistent repeating of the sample size is quite important, as the detector response ratio of the internal standard to the unknown components could change with large swings in sample size (usually greater than 0.5-μl). If simply doing weight or volume percent compositional work, sample size can also affect quantitative results, especially when there are a few major peaks and several smaller components. In general, the syringe techniques described previously in this chapter should be used to help insure good quantitative results via a good sample injection.

Question: A Hewlett-Packard 7620 chromatograph is equipped with an effluent splitter. What could cause hydrogen to back diffuse through an open end of the splitter into any other type of detector or reaction chamber attached?

Answer: These Hewlett-Packard chromatographs have a small pinhole in the flame detector connection opening that adds a small amount of hydrogen to the sample stream as it enters the detector to promote more complete combustion of the sample at the burner tip. If the column flow is turned off or drastically reduced (as would be the case for a capillary column), the splitter eluent line should be plugged-off to prevent hydrogen from backflushing through the splitter, and into the column oven or auxiliary detector being used. This is important, as over a period of time (inverse to the analytical column flow), an explosive hydrogen/air mixture could be attained, and an explosion would be waiting for anyone who came near the oven with a combustion source.

Question: Does the diameter of the opening in a septum retainer cap have any bearing on the life of the septum?

Answer: As a matter of fact, yes! Some openings are only slightly larger than a syringe needle, while others are as much as 1/4-inch in diameter. Contrary to common beliefs, the septum that is punctured in essentially the same place through the small retainer cap hole has much less chance of leaking than the one that is punctured in several places. When puncturing the same hole each time, there is little chance for sample backpressuring through the septum on vaporization, as the hole is filled with the syringe needle. Not so on septa that have many puncture holes. So, in general, the smaller the size of the opening in the septum retainer, the longer useful life the septum should have.

Question: How can leaks be located in a GC gas system without checking the fittings with a soap-film solution?

Answer: Leaks can be most readily detected by use of a helium leak detector. In the past, most leak detectors for helium were small, dedicated mass spectrometers that had to be wheeled from job to job via a cart or some other type of conveyance due to their heavy weight. Now available are small, light helium leak detectors based on the thermal conductivity principle of operation. They can either be used with AC line power to the air-sampling pump or DC battery power for field use. Such detectors are very sensitive and able to detect leaks as low as 10-6cc/second. The leak detector was discussed briefly earlier in this chapter.

Question: With certain model chromatographs, workers are unable to reproduce data to 1% repeatability using an internal standard. What could be wrong?

Answer: Generally, with an internal standard technique, 3% repeatability is considered good, so they may be operating as accurately as possible. If flow variations are changing detector sensitivity during a programmed-temperature run, the first thing to do is clean and leak-test the flow controller as described earlier in this chapter. If flow variations still persist, the controller should be replaced. In addition, if a thermal conductivity detector is used, variations in flow with changing column temperature will also cause considerable baseline drift not detectable with an FID. This certainly sounds like a flow-controller problem, and should be dealt with as such. If quantitation variations continue after being assured of correct operation of the flow controller, the chromatographer must look elsewhere for the problem, such as the injection system, the method of injection, small, virtual leaks, inconsistent operation of the detector itself, or simply poor peak area measurement as is discussed in Chapter 12.

Question: Generally, what is the easiest way to set up for on-column injection with units that have injection-port liners, such as a Hewlett-Packard 5750?

Answer: For 1/4-inch O.D. columns, simply remove the injection-port liner, and insert the front part of the column through the bulkhead fitting as if it were the liner itself. Remember to lower the heat of the injection port below the maximum column temperature to preserve the stationary phase at the column inlet. For 1/8-inch O.D. columns, it would probably be best to put a larger I.D. liner in place (5/32-inch I.D.) in the usual way, then drill out the 1/8-inch column adapter fitting and insert the 1/8-inch column through the entire length of the 5/32-inch I.D. liner, to the head of the inlet system, where on-column injections may then be carried out as for a 1/4-inch O.D. column.

Question: Of what use are backpressure gauges?

Answer: Only to see if extensive backpressuring occurs on vaporization of large quantities of light material (10 μl) or gaseous materials when they expand on entering the heated vaporizer region. This gives some idea on selecting correct sample sizes.

Question: What is the best injection method for getting reproducible samples of a material that sets up in the syringe?

Answer: Dissolve the material in a known amount of a suitable solvent, and then reproducibility inject known amounts of this solution using the "flush" technique as described when discussing injection procedures earlier in this chapter.

Question: Should the flow controller be followed by the rotameter or the reverse of this sequence?

Answer: In general, the rotameter precedes the flow controller as shown in Figure 1 of this chapter. However, if the rotameter is placed downstream of the controller, it gives a better reading of actual column flow, and thus a direct indication as to whether the flow controller is performing correctly.

Question: What is the possibility of determining and reproducibly sampling trace organic components in small volumes of air?

Answer: The determination of trace amounts of organics in air can be done quite readily, but only with some intermediate trapping step to concentrate the trace organic constituent. Thus, large known volumes of air must be pulled through the organic trap (such as activated charcoal) to concentrate the trace organic, and this concentrate displaced from the trapping material and analyzed by GC. Not enough air could be injected into a GC unit to determine trace amounts without grossly overloading the GC column.

Question: A 1.0 µl sample of a 2.5% solution of extremely polar and nonvolatile triglycerides in chloroform was injected in the usual manner at 360°C onto a 3% OV-1 on Gas Chrom Q column. Severely tailing peaks resulted. How could this tailing be overcome?

Answer: The solution was one of simply using the "solvent-flush" technique of syringe injection to wash residual triglycerides from the syringe needle instead of letting them dribble out while the syringe needle is being withdrawn from the vaporizer.

Question: A particular process stream contains H_2, H_2S, and C_1-C_{14} hydrocarbons. What is the best way to inject such a sample?

Answer: Probably via a heated gas-sampling valve. Since this is a gaseous stream, a gas sampling valve would be best suited for the analysis, with the heat supplied to keep from losing any of the heavier hydrocarbons by condensation. A valve temperature from 150-200°C would be sufficient.

Chapter 9

PYROLYTIC INLET SYSTEMS

INTRODUCTION

Pyrolysis is not yet used by the majority of chromatographers. The technique of pyrolysis gas chromatography (PGC) is a rapidly growing field of chemistry and its potential for use in the analysis of solids, polymers, and many organics warrants its discussion here. Primarily, a lack of interlaboratory reproducibility has been responsible for the lack of widespread use of PGC, but advances have been made and are being made in this area. Although we will not treat the subject in detail here, it is hoped that the discussion of this topic will aid the chromatographer engaged in PGC to maintain and troubleshoot his system.

It was not long after the introduction of gas chromatography that Davison, Slaney, and Wragg (1) combined GC with pyrolysis to form pyrolysis gas chromatography. Pyrolysis itself had been used as long ago as 1860 by Williams (2) to determine the structural unit of natural rubber. Essentially, pyrolysis is the process of heating a substance to high temperatures, thereby causing fracturing of the substance into lower molecular weight compounds. (In a very few instances, pyrolysis will result in larger molecules being formed, but these are of less importance.)

Introducing the pyrolysates (products of the pyrolysis) into the chromatograph results in a pyrogram. This pyrogram can provide means of identifying an unknown substance since the peaks are uniquely characteristic of the starting material (only stereoisomers will have identical pyrograms). Also, the pyrogram can provide information concerning the structure of the starting material, be it a pure compound, a mixture of compounds, or a polymer. PGC has found use in many fields such as criminology,[3] biology,[4] medicine,[5] polymers,[6] petroleum,[7] and many other areas of interest to chemistry.

REQUIREMENTS

To obtain this information from the pyrolysis, certain requirements must be met. As with any analytical technique, the results must first be repeatable. This, of course, means that all factors determining which products are formed must be held constant. Probably the most important factor affecting the distribution of the products formed is pyrolysis temperature. Figure 1 shows 3 pyrograms of polystyrene. In the first pyrogram, the pyrolysis temperature was 425°C, in the second pyrogram the pyrolysis was 825°C, and in the third it was 1025°C. The amount of product is drastically different. Because large samples are difficult to heat rapidly and uniformly, small thin film samples are preferred. Another factor affecting the product distribution is illustrated in Figure 9-2. This figure compares sample size to the amount of products produced from the pyrolysis of methyl octanoate (8). Longer periods of pyrolysis will result in greater portions of the starting material being pyrolyzed. At first glance, this might seem to be an advantage, but seldom can any additional information be gained past the first instant of pyrolysis, and the additional pyrolysis time may only serve to cloud the interpretation of the chromatogram.

Secondary reactions are important when PGC is used in structure elucidation. When determining the structure of the starting material (i.e., a polymer), the most information can be determined from the initial pyrolysis. Secondary reactions are the result of reactions between the pyrolysates themselves, or may be caused by reaction of the pyrolysates with the interior walls of the pyrolysis chamber. While the reaction vessel may not necessarily react directly with the pyrolysates, it may act as a catalyst. Large dead volumes in the pyrolyzer promote secondary reactions. Steps should be taken to limit secondary reactions. Reaction vessels are made of non-catalytic materials such as quartz, glass and gold. In cases where more reactive materials must be used (such as platinum, iron, cobalt, and nickel) precautions should be taken to insure that these materials are not entering into reactions with the pyrolysates. The sample should also reach the chosen pyrolysis temperature as rapidly as possible. Rapid temperature rise times (TRT) may

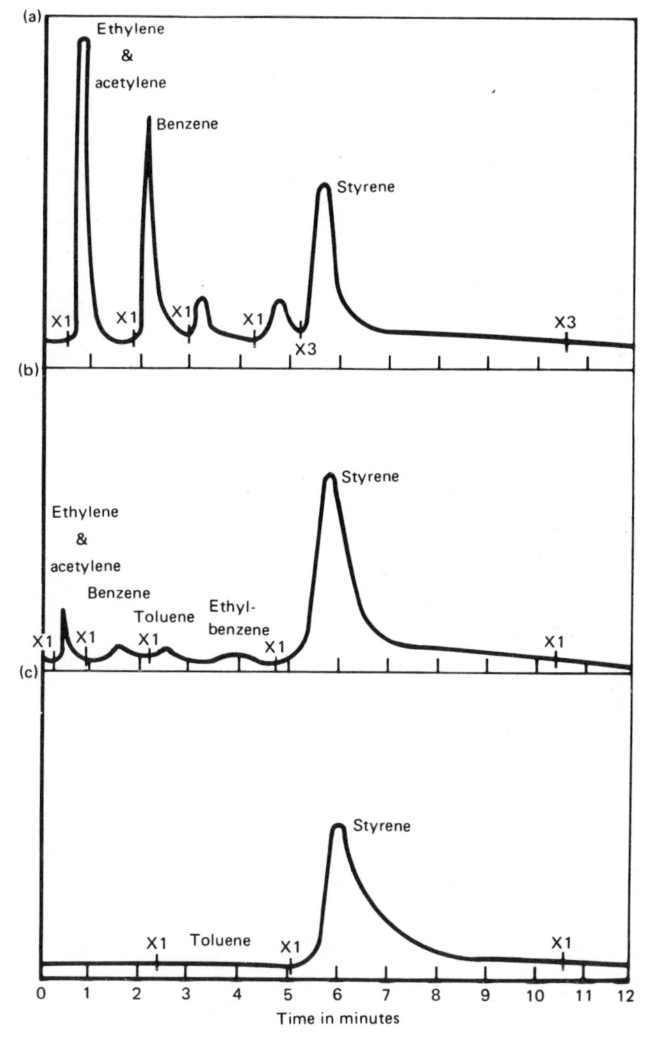

Fig. 9-1. Pyrograms of polystyrene at various pyrolysis temperatures. (A) 452°C, (B) 825°C, (C) 1025°C. Column: Apiezon L; column temperature; 140°C; flow rate; 60ml./min. Attenuation scale indicated by number in Figure. (Courtesy of Anal. Chem.) *(10).*

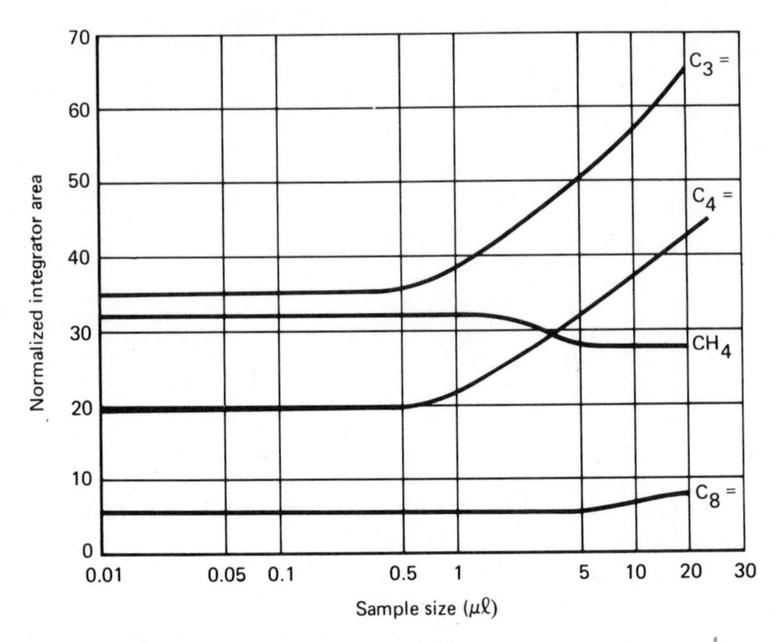

Fig. 9-2. Variation in the relative yields of CH_4, C_3H_6, C_4H_8, and C_8H_{16} as a function of sample size from pyrolyzed decanoate. (8) (Courtesy of Anal. Chem.)

eliminate pyrolysis below the chosen pyrolysis temperature.

To summarize, a good pyrolysis unit should provide easily repeatable procedures. It should also exhibit a fast TRT (less than one second); and, whenever possible, non-reactive and non-catalytic surfaces should be utilized in any area with which the pyrolysate might come into contact. The sample should be applied in the form of a thin film or, where possible, in the form of a vapor.

Many of these requirements have been met by today's pyrolysis apparatus. Within a given pyrolysis system, a high degree of repeatability can be established. Unfortunately, due to the number of different pyrolyzers, reproducibility between different laboratories has been difficult to achieve. Essentially what is needed is a set of standard pyrolysis conditions, but little progress has been made to this end so far. Nonetheless, PGC has gained wide acceptance as an analytical instrument in several fields.

INSTRUMENTATION

Early apparatus in PGC consisted of heated reaction vessels from which the pyrolysates were collected by means of a syringe and injected into the chromatograph. This method was sometimes augmented by liquid nitrogen cooling of pyrolysates to insure collection of low molecular weight compounds (9) (10). The method is seldom used today, having been replaced by much more reliable and informative methods.

Current methods utilize direct introduction of the pyrolysates onto the GC column. The carrier gas flows continuously through the pyrolyzer unit and into the column. Pyrolyzers are classified as either continuous mode (if the sample is introduced into a pyrolyzer pre-heated to the pyrolysis temperature) or pulse mode (if the sample is introduced into a cold pyrolyzer which is then brought rapidly to the pyrolyzer temperature). Each type of pyrolyzer has characteristic advantages and disadvantages which determine its use.

Continuous mode pyrolyzers are the simplest in construction. The first of this type to be used was the boat pyrolyzer. The sample is placed in a sample "boat" which is resting in a cool zone within the carrier gas stream. For pyrolysis, the boat is moved mechanically (some units use a magnet) into the pre-heated pyrolysis zone and the pyrolysates are swept into the column immediately. Boat type pyrolyzers are rarely used today, primarily because they are susceptible to secondary reactions caused by large dead volumes and a slow heat-up time.

Closely related to the boat pyrolyzer, the vapor phase tubular reactor is still used frequently today. As diagrammed in Figure 9-3, pre-heated carrier gas flows into a heated injection port. Sample is injected through the injection port into the vaporizer unit where it is vaporized, but not pyrolyzed. The vapors then enter the pyrolyzer section which has been maintained at the pyrolysis temperature. The pyrolyzer often consists of a length of gold tube which limits catalytic reactions and prevents residual sample acumulating which might interfere with subsequent analysis. The pyrolysates are swept immediately into the GC system, traveling through tubing which is heated to prevent condensation.

Vapor phase tubular reactors afford many advantages over previous methods. No secondary reactions can occur, and the results are very repeatable. These units are very simple in design, and the pyrolysis temperature can be set at any position over a wide range of temperatures. Another advantage, especially when determining structure, is that the pyrolysates can be predicted with a fair degree of accuracy on the basis of the Kossiakoff and Rice theories of free radical degradation (11). These pyrolysis units are also relatively inexpensive when compared to other types used today.

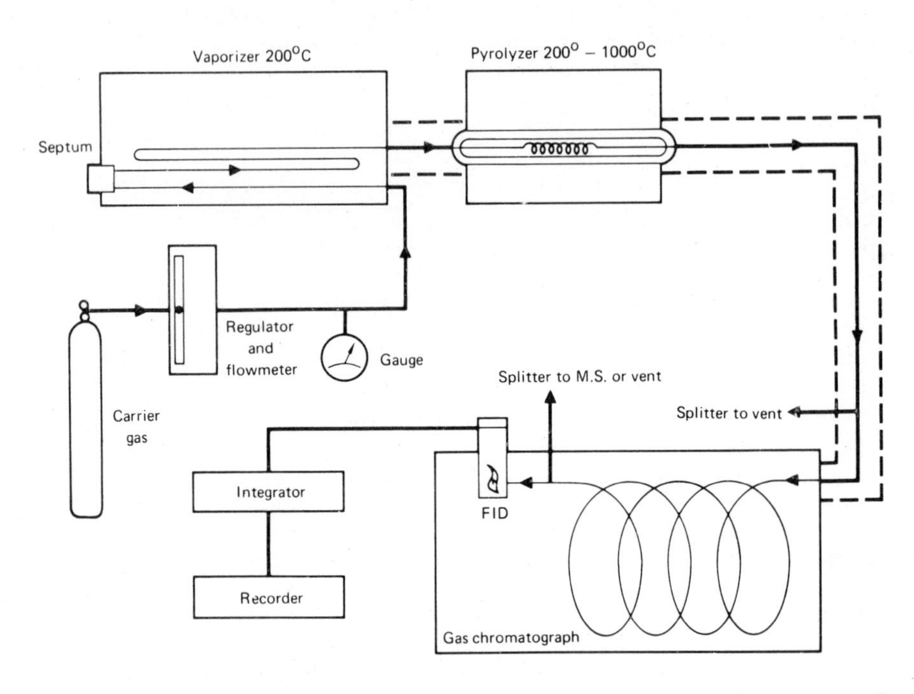

Fig. 9-3. Vapor-phase Pyrolysis/GC apparatus. (Courtesy of Anal. Chem.) *(17)*

Unfortunately, vapor phase pyrolyzers are limited seriously in that they can only be used with readily volatile substances. They cannot be used, therefore, with solids and low vapor pressure liquids above C_{20}. Since the bulk of PGC work done today is in this area, other types of pyrolyzers have been developed to fit those needs.

The first of the pulse mode pyrolyzers to be used was the resistance heated filament pyrolyzer (12-15). The filament is enclosed in a glass tube through which carrier gas is flowing. Preferably, the sample is applied by dipping the filament in a solution of the sample, then evaporating the solvent. This method provides a thin uniform film of sample. Alternatively, a solid sample can be positioned on the filament, but it is difficult to apply small samples in this manner. With the sample in place, current is applied to the wire causing it to heat up and the sample is pyrolyzed.

These fairly inexpensive and widely used units can be heated over a wide temperature range. Resistance heated filament pyrolyzers exhibit some drawbacks. Slow and sometimes difficult to reproduce temperature rise times can result in poor reproducibility. The slow TRT also contributes to products characteristic of lower pyrolysis temperatures, making the pyrograms difficult to interpret. Recently however, much faster TRT's have been achieved by capacitive boosted filaments (16, 17) (see Figure 9-4). The secondary reactions with these units are nearly nonexistent and repeatability is enhanced accordingly. These units are not commercially available which prohibits their wide acceptance presently, but they will probably reach major importance in the future.

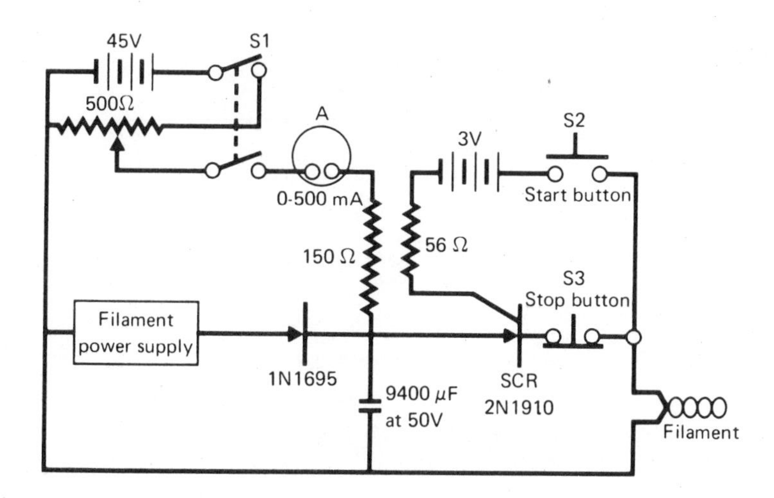

Fig. 9-4. Schematic of the capacitor discharge circuit for rapid excitation of filament pyrolyzers (18). (Courtesy of Anal. Chem.)

A second type of pulse mode pyrolyzer is the Curie Point Pyrolyzer. Developed in the early 1960's by Simon and his co-workers (18), these pyrolyzers utilize an interesting property of ferromagnetic materials. When exposed to a radio frequency (RF) field, a ferromagnetic wire will heat rapidly. The wire attains a specific temperature (called the Curie Point) depending on the wire alloy compositions and maintains that temperature until the RF field is turned off. Wires are available which will stabilize at a number of temperatures between 300 and 1000°C. Some Curie Point wires and alloy composition are shown in Table 1. A typical Curie Point apparatus is shown in Figure 9-5.

Several advantages are evident for Curie Point pyrolyzers. They can be used with solid soluble materials, liquids which can adhere to the ferromagnetic wire, and some crystals and powders. The pyrolysis temperature is determined precisely and reproducibly by the Curie Point of the

TABLE 9-1

Ribbon probe and platinum coil

Temperature rise time to 600°C	10msec (ribbon)
Temperature range	Up to 1000°C
Availability	Commercial
Repeatability	Good
Disadvantage	If solid material falls off use quartz tube in coil.
Advantages	1. Excellent rise time. 2. Wide temperature range. 3. Can be purchased. 4. Good repeatability.

wire. Temperature rise times of these pyrolyzers are fast, though it depends somewhat on the RF generating unit being used. These factors contribute the high degree of repeatability which can be obtained with Curie Point pyrolysis. The simplicity of their operation indicates that they may provide some answer to the problem of repeatability between laboratories.

Curie Point pyrolyzers are not without drawbacks. Some solid materials are difficult to apply to the wires. While secondary reactions are few, they can occur in some instances. Pyrolysis temperatures are not continuously variable as they are with other types of pyrolyzers, being limited to the Curie Points of the available wires. Table 9-2 shows the variations in Curie Point Pyrolyzers.

Several other types of pyrolyzers have been reported. Most of these are for limited use and will not be discussed here. Many, however, are included in Table 3, which compares the characteristics of the various pyrolyzers.

Ribbon Pyrolyzers-Several ribbon pyrolyzers have been examined, and at least one is commercially available (9). Ribbons have a larger surface area than wire type pyrolyzers and, thereby, thermally fragment more sample. Early ribbon designs were made of chemically reactive nickel metal which promoted chemical decomposition in competition with thermal decomposition. One commercial ribbon (Barber-Colman) is designed with a six port valve which is subject to leaks

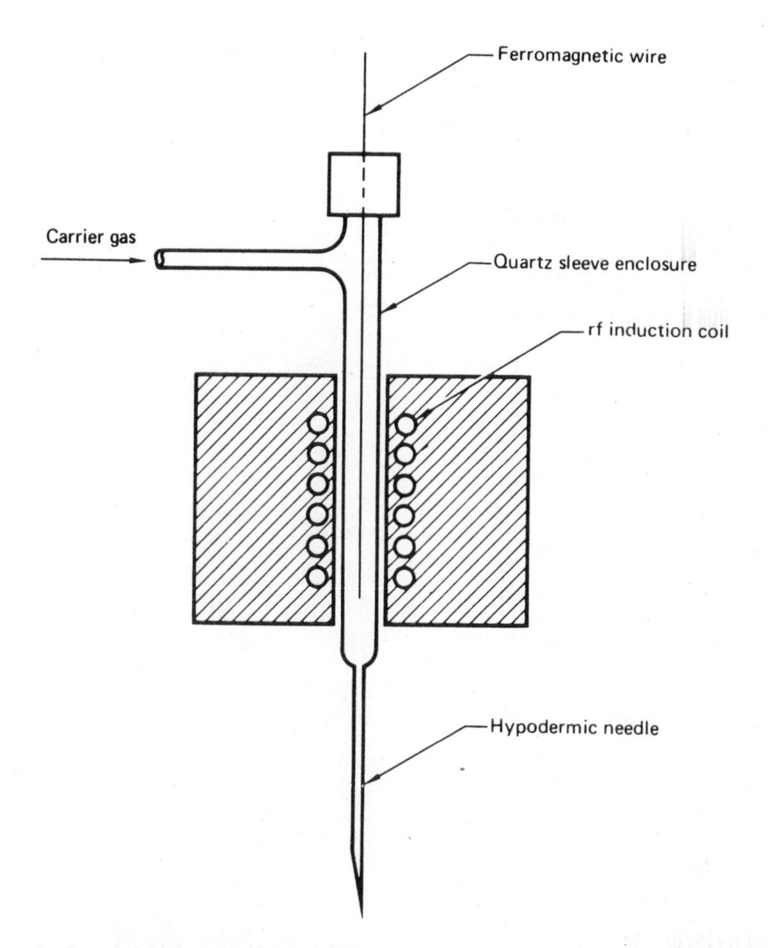

Fig. 9-5. Cross-section diagram of a Curie Point pyrolyzer (18). (Courtesy of Anal. Chem.)

when heated. The TRT of this unit varies from 2 to 10 sec. depending on the sample size and the mass of the ribbon. Recently, Martin *et al.* (21) described a platinum micro-ribbon which measures 3.8 cm long, 1 mm wide and 0.0005 inches thick which has a TRT of 8 to 17 msec. The power supply developed for this micro-ribbon is an adjustable wheatstone bridge circuit which controls both the final temperature and the rise time to that temperature. Problems sometime occur when fast TRT's are applied to some designs as the ribbon tends to "flip" portions of solid powder samples off the micro-ribbon. In such cases, the slower TRT's yield more repeatable sample sizes. This is very common with such samples as polyimides and Teflon (R) which do not dissolve in most solvents and are, therefore,

TABLE 9-2

Data for oscillators used in Curie Point pyrolyzers

Oscillator	Frequency of Oscillator [KHz]	Power Output [W]	Applied Magnetic Field [Oe]	Rise Time of Iron Wire (0.5 mm diam.) to Curie Point [sec]
Fischer	1200	1,500	382	0.08
Packard Model 891	?	90	?	0.25
Pye Unicam (Cat. No. 12 558)	550	30	?	ca. 2
Japan Analytical Industry Co., Ltd., Type JHP-ZA Curie Point GC Pyrolyzer	400-500	150	?	?
ETH (Simon)*	480	1,500	1,170	0.03

*Swiss Federal Institute of Technology, Jurich, Switzerland.

TABLE 9-3

Comparison of Curie Point Wires (6)

Manufacturer's specification °C	TGA °C***	Curie Point Range °C	Manufacturer's per cent Composition Fe-Ni-Co	X-ray fluorescence Fi-Bi-Co****
358*	352	±2	0-100-0	0-100-0
480*	474	±2	52-48 -0	53.5-46.5-0
510*	482	±2	49-51 -0	50.6-49.4-0
600**	597	±12	42-42 -0	42-42-16
610*	601	±4	30-70 -0	29.2-70.8-0
700**	707	±12	33-33-33	33-33-33
770*	782	±4	100-0-0	100-0-0
980*	985	±5	0-60-40	39-1-60

*Philips Electronic Instruments, Mt. Vernon, N.Y.

**Fischer Labs (Varian-Aerograph, Walnut Creek, Calif.)

***Thermal Gravimetric Analyses (Magnetic transition point, corrected). Perkin-Elmer TGS-1.

***Avg. of three X-ray fluorescence analysis from Siemens Crystalloflex IV X-Ray Generator with vacuum tunnel spectrometer.

(Courtesy of *Analytical Chemistry*.)

difficult to apply to wire or ribbons. To circumvent this sampling problem, a different probe configuration is supplied with each unit.

This second probe consists of a platinum wire coiled around a very thin quartz tube. The sample in any form (liquid, film or powder) is contained in a thin quartz tube with "quartz wool." The platinum coil around the quartz tube has a resistance of 0.25 ohms. Each probe is supplied with a final temperature calibration chart. The temperature rise time of this quartz probe is somewhat slower than the ribbon probe. However, the rise time is less than 100 msec to 700°C. Also this probe is very useful for sampling materials which do not dissolve in solvents. [There is a boat pyrolyzer (Fisher Scientific) commercially available for the thermal decomposition of "powder" or "gel" samples which do not dissolve in most solvents. The temperature rise times are in the range of 200°C per second. Obviously, most of the sample within such a boat unit will decompose long before the final temperature is reached.] Table 3 gives the characteristics of the ribbon probe and platinum coil pyrolyzers.

RIBBON CLEANING

It is important to have a "clean" ribbon prior to applying any sample. Cleaning is done in a short time by raising the temperature of the ribbon to 600°C or higher *in air* for at least five seconds. Then immediately readjust the time and temperature dials to the desired pyrolysis conditions.

TEMPERATURE RISE TIME

The effect of temperature rise time is shown in Figure 9-6. The upper pyrogram of a thin film of polyvinyl chloride was obtained with a TRT of 1.5 sec and the lower pyrogram of the same polymer with a TRT of 15 msec (10). Note that the production of the lower molecular weight materials was enhanced with the 15 msec TRT since the polymer was pyrolyzed closer to the true temperature of 700°C and, therefore, allowed less time for recombination of lower fragments to produce the heavier products.

PYROLYZER SAMPLING METHODS

One of the following procedures should be carried out by applying samples to the ribbon.

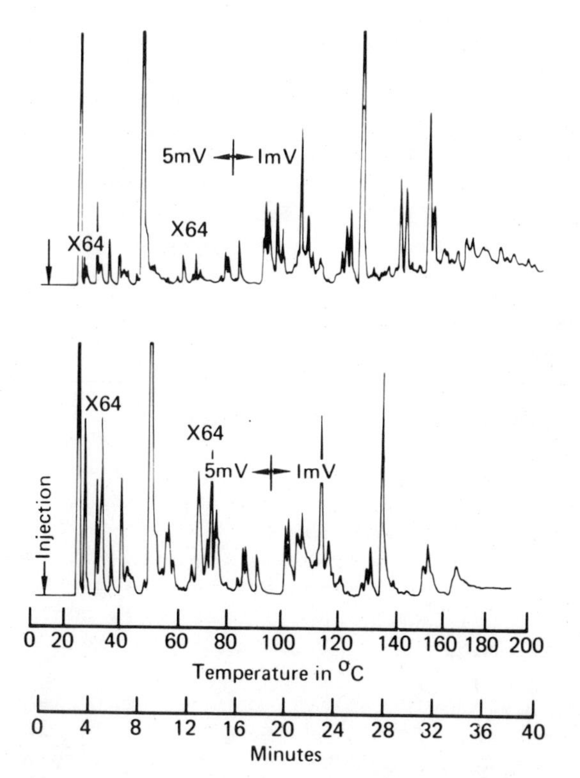

Fig. 9-6. The upper pyrogram illustrates pattern of Tygon at 700°C with a temperature rise time of about 1.5 sec. The lower pyrogram illustrates the pyrolysis pattern of Tygon at 700°C with a temperature rise time of 15 msec.

1. *Soluble Samples.* The best results are obtained by dissolving the polymer in a low boiling point solution such as ethylacetate, methanol, acetone, benzene or chloroform, to form a 10 percent-by-weight solution. Sample can then be drawn up into a one to ten microliter syringe from the 10 percent sample solution mentioned above. It is important to rest the pyro-probe-ribbon on a solid secure surface, such as a clean ash tray on a tabletop, since it is easier to apply the solvent/sample mixture to the ribbon, the probe is carefully placed into the injection port of the gas chromatograph. It is important that the injection port temperature be below 200°C since some samples will begin to decompose thermally at 200°C. The injection port temperature should be maintained to at least 100°C so that the solvent will quickly evaporate. Of course, the analyst must wait until the solvent has been swept through the chromatograph

prior to activating the pyrolyze button.

2. *Solid, ungrindable samples,* e.g., some rubbers are usually cooled to liquid N_2 temperatures and filed with a clean file or cut with a clean razor blade. Place this sample piece carefully in the center of the Pt. foil and carefully slide into the injection port. It is desirable to turn off the carrier gas during this injector loading step.

3. *Solid, grindable samples* do not need to be cooled to liquid N_2 temperatures prior to cutting filing, or grinding.

TROUBLESHOOTING

The chromatographic system used in PGC is essentially the same as that used in conventional GC and requires no comment. Troubleshooting of the PGC system requires attention to special problems which may arise from the pyrolysis unit. The tables which follow are divided into three parts. Part I is concerned with problems involved with pulse mode pyrolyzers (Curie Point, Electrical Discharge, Capacitive Boosted Discharge, Laser, etc.); Part II is concerned with problems which might arise from continuous mode pyrolyzers; Part III deals with problems encountered with either type.

Two points should be kept in mind when maintaining pyrolysis systems. First, all sample transfer lines should be kept heated at all points. This prevents condensation of pyrolysates before they can enter the column. Another important point is the frequent occurrence of leaks. Since most of the system is maintained at high temperatures for prolonged periods, oxidation and deterioration of seals is likely. At extremely high temperatures it may be necessary to replace septums every four hours. Metal tubing fittings can develop leaks also at these high temperatures.

PART I

Problems encountered with pulse mode pyrolyzers

Symptom of Trouble	Possible Cause	Remedy and/or Check
Poor reproducibility of relative peak areas from programs.	1. Variation in final temperature.	Weld Microthermocouple to filament and measure temperature with an oscilloscope (20).
		Measure magnetic transition point via TGA of Curie Point wires (20).
	2. Variation in temperature rise time.	(Same as above.)
	3. Variation in carrier gas pressure and velocity.	Attach an accurate pressure gauge and a reliable flow controller.
	4. Variation in coil or geometry.	Figure "8" better (15).
	5. Variation in sampling techniques.	Sample from suitable solvent 50 gm/liter (17).
		Sprinkle sample in powder state onto wire at 100°C.

PART I - Cont'd.

Problems encountered with pulse mode pyrolyzers

Symptom of Trouble	Possible Cause	Remedy and/or Check
Poor reproducibility of relative peak areas from programs, cont'd.	6. Variation in sample thickness.	200-300 Angstrom units is considered ideal (17).
	7. Variation in duration of pyrolysis temperature	Short durations of 1-3 sec. ideal (6). (Measure as in A.)
	8. Variations in pyrolyzer assembly temperature.	Pyrolysis product condensation can be reduced by maintaining pyrolyzer temp. from 100-200°C.
	9. Variations in pyrolyzer dead volumes.	Large dead volumes promote secondary reactions.

PART II

Problems encountered with continuous mode pyrolyzers

Symptom of Trouble	Possible Cause	Remedy and/or Check
Poor reproducibility of relative peak area from pyrograms.	1. Variation in constant temperature.	Potentiometer temp. check over a period of time.
		If a good temp. controller is not available for the reaction, use a constant voltage transformer.
	2. Variation in temperature profile of reaction.	Plot temperature profile of reactor. Try to keep sample in a constant temperature profile.
	3. Sample size varying.	Check per cent decomposition.
	4. Variation in carrier gas flow rate pressures.	Attach a good pressure gauge and flow controller. Measure and record pressure and flow rate.
	5. Catalytic reaction of sample with reactor wall.	Check inside wall of reactor for evidence of reaction. Use gold or gold plated surface if possible.

PART III

Problems encountered with both pulse and continuous mode pyrolyzers

Symptom of Trouble	Possible Cause	Remedy and/or Check
Low pyrolysis product recoveries and/or baseline drift.	Leak in carrier gas system in pyrolyzer, column and/or detector.	Leak check system at regular intervals.
Pyrolysis product distributions more complex than anticipated.	Catalytic Rx occurring with pyrolyzer and/or column system.	Clean pyrolyzer and transfer lines and change reactive surfaces by coating with teflon or gold plating.
Peak broadening.	1. Large dead volumes.	A. Reduce dead volumes.
	2. Transfer lines-pyrolyzer to GC and/or GC to detector too cool.	B. Add thermo-couplers and/or heaters to check and maintain temperatures.
High molecular weight products do not elute from pyrolyzer.	Same as peak broadening causes.	Same as peak broadening causes.

Comparison of performance of various pyrolyzers

Characteristics Method	Sample Size	Ease of Applying Liquid and Solid Samples	Temperature Range	Pyrolysis Element Temperature Control	Catalytic and/or Secondary Reactions with Pyrolyzer
Conventional filament and ribbon (12)	10-1000 µg	1. Soluble in a solvent.	Up to 1200°C	Continuously variable	Yes
Curie Point high power (1500 watts) (18)	10-50 µg	1. Soluble in a sol- vent. 2. Some crystals and powders (heat wire to 100°C).	Up to 985°C	Limited to 352, 400, 482, 597, 601, 500, 707, 782, 985°C	Yes
Curie Point low power (30-100 watts) (6)	10-50 µg	1. Soluble in a solvent. 2. Some crystals and powders (heat wire to 100°C).	Up to 985°C	Limited to 352, 400, 482, 597, 601, 500 707,782, 985°C	Yes
Capacitive boosted fila- ment (17)	5-10 µg	1. Soluble in a solvent. 2. Heat wire to 100°C.	Up to 900°C	Continuously variable	No
Tube reactor and boat (15)	1-5000 µg	All forms liquids solids	Up to 1500°C	Continuously variable	Slight
Vapor phase (8)	.001-10 µg	Volatile materials only.	Up to 800°C	Continuously variable	No
Laser (19)	500 µg or larger	Dark sample or media such as carbon.	Estimated 1200-1500°K	Not controlled	Yes with graphite
Electric discharge (15)	200 µg	All forms		N/A	Slight

Characteristic Method	Temperature Rise Time (TRT)	Available on the Market	Repeatability	Disadvantages	Advantages
Conventional filament and ribbon (12)	2-10 sec	Yes	Fair	1. Slow rise time. 2. Aging of metal. 3. Solid meterial hard to apply.	1. Most widely used. 2. Wide temp. range. 3. Purchased or built. 4. Simple to apply sample.
Curie Point high power (1000 watts) (18)	40-120 Milli-seconds	Yes	Good	1. Solid material hard to apply. 2. Pyrolysis chamber must be heated.	1. Fast rise time. 2. Accurate temp. 3. Good repeatability. 4. Can be purchased.
Curie Point low power (30-100 watts) (6)	0.5-2 sec	Yes	Fair	1. Solid material hard to apply. 2. Pyrolysis chamber must be heated. 3. Slow rise time.	1. Accurate temp. 2. Fair repeatability. 3. Can be purchased.
Capacitive boosted filament (17)	15 milli-seconds	No	Good	1. Solid material must be soluble. 2. Must use small sample sizes. 3. Cannot be purchased.	1. Fast rise time. 2. Wide temp. range. 3. Good repeatability.
Tube reactor and boat	30-50 sec.	Yes	Fair	1. Slow rise time. 2. Requires large samples. 3. Does not lend itself to capillary columns.	1. Any material liquid or solid. 2. Widest temp. range. 3. Wide sample size range. 4. Can be purchased. 5. Results can be compared to kinetic studies.
Vapor phase	N/A	Yes	Good	1. Only volatile materials.	1. Good repeatability. 2. Wide temp. range. 3. Can be correlated with theory.

Characteristic Method	Temperature Rise Time (TRT)	Available on the Market	Repeatability	Disadvantages	Advantages
Laser	Estimated 10 sec	No	Good (Blue glass)	1. Must use colored agent to pyrolyze. 2. Complicated. 3. Large sample size. 4. Equilibrium temp. higher than other pyrolyzers. 5. Laboratory hazards. 6. Cannot be purchased. 7. Complex fragmentation process.	1. Pyrograms may be simple. 2. Superior rise time. 3. Not necessarily thermal degradation.
Electric discharge	50-100 sec	Yes	Poor	1. Slow rise time. 2. Large sample required. 3. Poor repeatability.	1. Any material liquid or solid. 2. Wide temp. range.

REFERENCES

1. W. H. T. Davison, S. Slaney and A. L. Wragg, *Chem. Ind.* (London) 1356 (1954).

2. C. G. Williams, *Phil. Trans.*, *150*, 241 (1860)

3. P. L. Kirk, *J. Gas Chromatog.*, *5*, 11 (1967).

4. P. M. Adhikary and R. A. Harkness, *Anal. Chem.*, *41*, 74 (1971).

5. E. Reiner and G. P. Kubica, *Am. Rev. Respirat Diseases*, *99*, 42 (1969).

6. M. T. Jackson Jr. and J. Q. Walker, *Anal. Chem.*, *43*, 74 (1971).

7. S. G. Perry in *Advances in Chromatography*, Vol. 7, J. C. Giddings and R. A. Keller, eds., pp. 221-240, Marcel Dekker, New York, 1968.

8. W. D. Dencker and C. J. Wolf, *J. Chrom. Sci.*, *8*, 534, (1970).

9. H. Feurberg and H. Z. Weigel, *Anal. Chem.*, *199* (2), 121 (1963).

10. J. Strassburger, G. M. Brauer, M. Tyron and A. F. Forziatti, *Anal. Chem.*, *32*, 454 (1960).

11. A. Kossiakoff and F. O. Rice, *J. Am. Chem. Soc.*, *65*, 590 (1943).

12. K. Ettre and P. F. Varidi, *Anal. Chem.*, *34*, 1543 (1963).

13. S. Tsuge, T. Okumoto, and T. Takeuchi, *Kogyo Kagaku Zasshi*, *71*, 1634 (1968).

14. R. L. Levy, *J. Chromatog.*, *34*, 249 (1968).

15. R. L. Levy, *Chromatog. Rev.*, *8*, 59 (1966).

16. R. L. Levy, *J. Gas Chromatog.*, *5*, 107 (1967).

17. R. L. Levy, D. L. Fanter, and C. J. Wolf, *Anal. Chem.*, *44*, 38 (1972).

18. W. Simon and H. Giacoggo, *Chem. Ing. Tech., 37,* 709 (1965).

19. D. L. Fanter, R. L. Levy and C. J. Wolf, *Ibid., 44,* 43 (1972).

20. R. L. Levy and D. L. Fanter, *Anal. Chem., 41,* 1465 (1969).

21. A. J. Martin, S. F. Sarner, and E. J. Levy, Pittsburgh Conference on Analytical Chemistry and Applied Spectroscope, Cleveland, Ohio (March, 1971).

22. J. Q. Walker, *Chromatographia, 5,* 547 (1972).

Chapter 10

GAS CHROMATOGRAPHIC COLUMN OVENS AND TEMPERATURE CONTROLLERS

The purpose of this chapter is to present sufficient information to allow a worker in gas chromatography to recognize and correct malfunctions in analytical columns, the column thermostat (or oven), the column oven temperature controller or programmer, and the sample transfer lines running from the inlet to the detector. The column system is the heart of the chromatograph. If it is not functioning properly the finest inlet and detector systems are of little practical use. The desired function of each component will be described along with routine maintenance procedures and actual laboratory maintenance problems.

The nucleus of all GC systems lies in the analytical column where the separation of the components in a mixture actually occurs. The right column in GC is one that has the proper size and stationary phase. This column must be operated under the optimum conditions that will separate the components desired. To select the column liquid stationary phase, "like dissolves like" is a good general rule to follow. Thus, nonpolar liquid phases are best for separating nonpolar mixtures, such as paraffinic hydrocarbons, while polar stationary phases generally provide optimum separation of polar compounds, such as alcohols or phenols. Table 10-1 shows some examples of stationary (liquid) phases, their molecular structure, approximate useful temperature range, and the classes of materials they would separate (1).

TABLE 10-1

Stationary phases for analysis of sample types shown

Liquid Films (in order of increasing polarity):

For saturated hydrocarbons and other non-polar compounds:

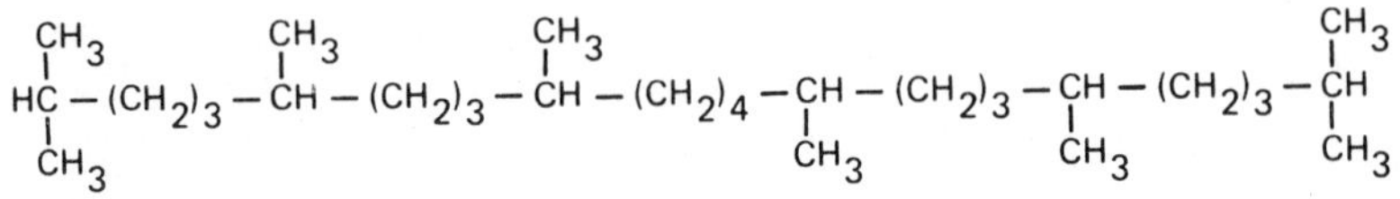

$$\underset{\displaystyle CH_3}{\overset{\displaystyle CH_3}{HC}}-(CH_2)_3-\overset{\displaystyle CH_3}{CH}-(CH_2)_3-\overset{\displaystyle CH_3}{CH}-(CH_2)_4-\underset{\displaystyle CH_3}{CH}-(CH_2)_3-\underset{\displaystyle CH_3}{CH}-(CH_2)_3-\underset{\displaystyle CH_3}{\overset{\displaystyle CH_3}{CH}}$$

0 to 100°C

Squalane

For unsaturated hydrocarbons and other semi-polar compounds:

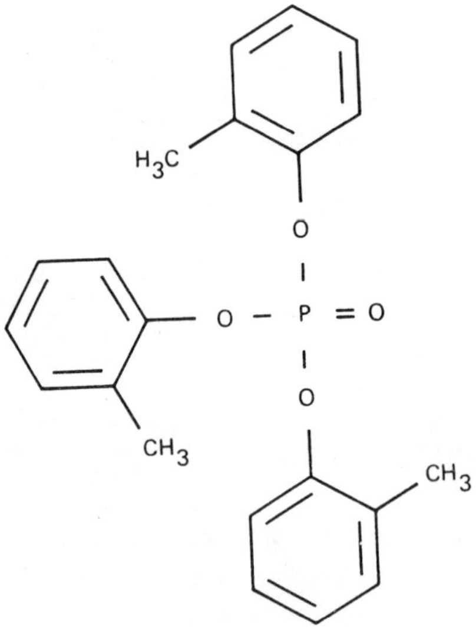

H_3C CH_3 CH_3

O – P = O

Tri cresylphosphate 20 to 125°C

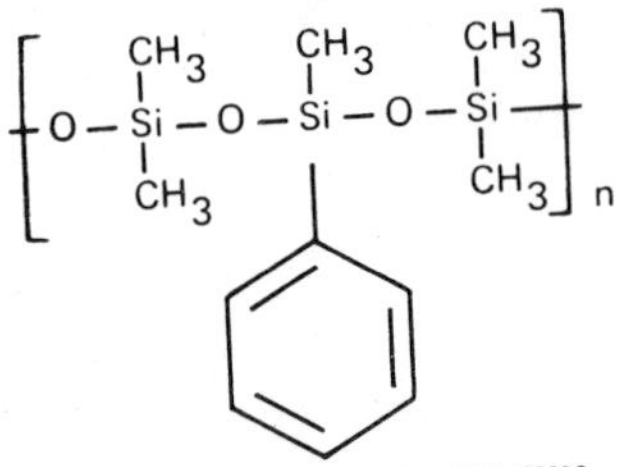

$$\left[-O-\underset{\displaystyle CH_3}{\overset{\displaystyle CH_3}{Si}}-O-\underset{\displaystyle C_6H_5}{\overset{\displaystyle CH_3}{Si}}-O-\underset{\displaystyle CH_3}{\overset{\displaystyle CH_3}{Si}}-\right]_n$$

SE - 52 50 to 300°C

Dow - 710 } −15 to 250°C

Dow - 550

TABLE 10-1 - Cont'd.

Stationary phases for analysis of sample types shown

For alcohols, esters, ketones and acetates:

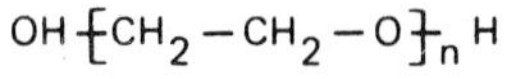

$$OH{-}[CH_2-CH_2-O]_n{-}H$$

Carbowax
20M 60 to 220°C
400 10 to 100°C

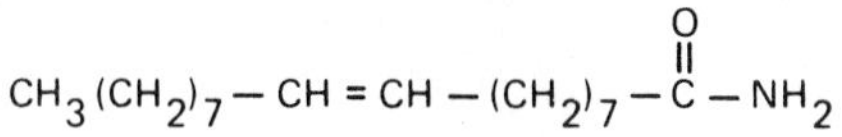

$$CH_3(CH_2)_7-CH=CH-(CH_2)_7-\overset{\overset{\displaystyle O}{\|}}{C}-NH_2$$

Hallcomid M 180L −8 to 150°C

For fatty acid methyl esters:

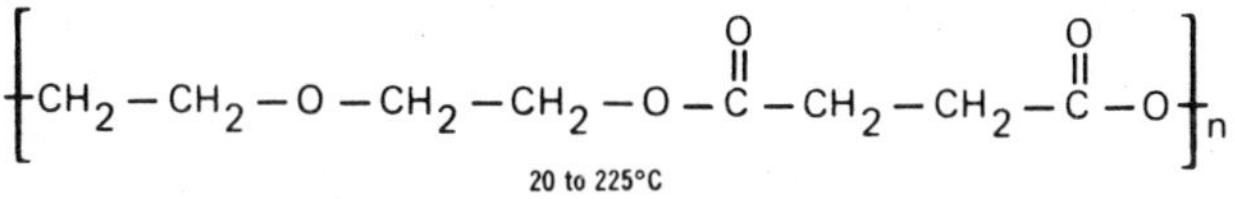

$$\left[-CH_2-CH_2-O-CH_2-CH_2-O-\overset{\overset{\displaystyle O}{\|}}{C}-CH_2-CH_2-\overset{\overset{\displaystyle O}{\|}}{C}-O-\right]_n$$

20 to 225°C
Diethyleneglycol succinate
Highly polar

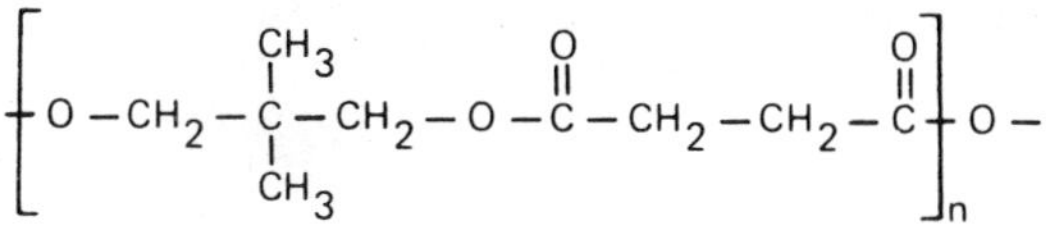

$$\left[-O-CH_2-\underset{\underset{\displaystyle CH_3}{|}}{\overset{\overset{\displaystyle CH_3}{|}}{C}}-CH_2-O-\overset{\overset{\displaystyle O}{\|}}{C}-CH_2-CH_2-\overset{\overset{\displaystyle O}{\|}}{C}-\right]_n O-$$

50 to 225°C
Neopentyl glycol succinate
Slightly polar

TABLE 10-1 - Cont'd.

Stationary phases for analysis of sample types shown

For nitrogen compounds:

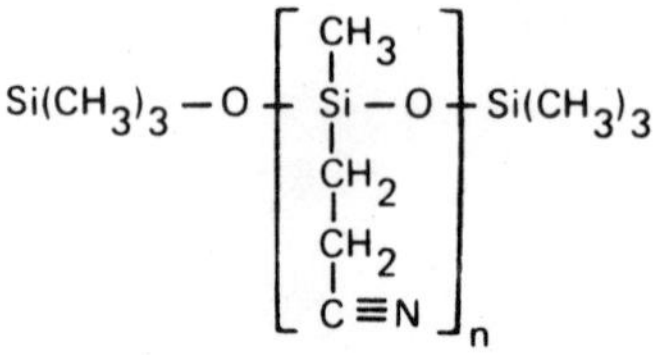

$$Si(CH_3)_3-O-\left[\begin{array}{c} CH_3 \\ | \\ Si-O \\ | \\ CH_2 \\ | \\ CH_2 \\ | \\ C\equiv N \end{array}\right]_n-Si(CH_3)_3$$

XF - 1150 XF - 1112
XE - 60 XF - 1125
0 to 240°C

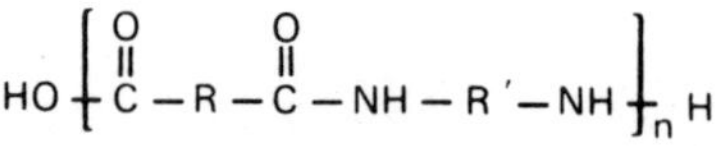

$$HO\left[\overset{O}{\overset{\|}{C}}-R-\overset{O}{\overset{\|}{C}}-NH-R'-NH\right]_n H$$

Versamid

190 to 275°C

For halogen compounds (incl. pestacides), freons and alkaloids:

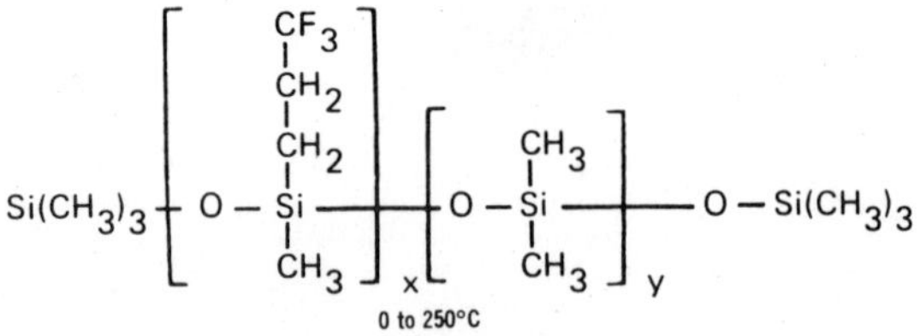

$$Si(CH_3)_3\left[O-\begin{array}{c} CF_3 \\ | \\ CH_2 \\ | \\ CH_2 \\ | \\ Si \\ | \\ CH_3 \end{array}\right]_x\left[O-\begin{array}{c} CH_3 \\ | \\ Si \\ | \\ CH_3 \end{array}\right]_y-O-Si(CH_3)_3$$

0 to 250°C

QF-1 (FS - 1265)

$$CF_3\left[CF_2\right]_n CG_3 \qquad Cl\left[CF_2CFCl\right]_n Cl$$

Kel F oil No. 10, No. 3 50 to 150°C Kel F grease

TABLE 10-1 - Cont'd.

Stationary phases for analysis of sample types shown

Solid Support

For glycols and gases (CO_2, H_2S, CS_2, N_2O, and NO):

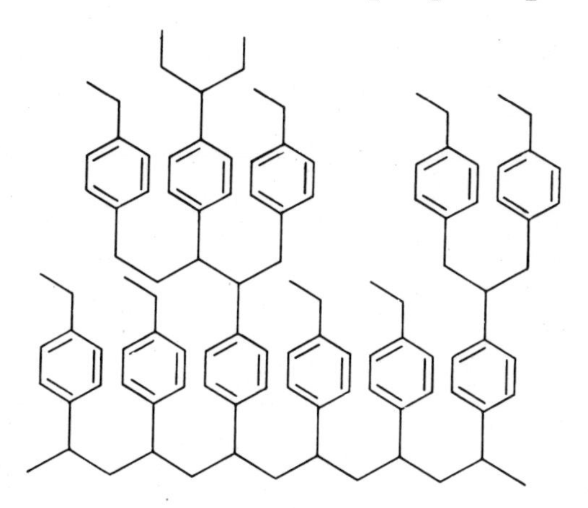

Porapak −20 to 225°C

The column dimensions should also be optimized for capacity (sample size) and speed of analysis. Table 10-2 gives the features and characteristics of typical analytical and preparative-scale columns (2). The capillary, 0.1158 cm and 0.317 cm outer diameter columns are used for greatest efficiency. These columns have small inner diameters, use small particle-size solid supports, or thin liquid phase films and require very small sample sizes (1/100mg.). The 0.635 cm outer diameter columns sacrifice some efficiency for capacity in semi-preparative GC work. The 0.952 cm outer diameter columns are used when capacity, not difficult separations, is required.

Meticulous attention to detail in preparing columns can ensure good operation. Many chromatographers do not take enough time or care in column preparation, and as a result, spend considerable amounts of time and money trying to achieve difficult separations using a mediocre column. Contrary to conjecture, column preparation is indeed a science. Duplicate columns prepared from the same set of ingredients should indeed be very similar in properties. This

TABLE 10-2

Comparison of column sizes

Parameter ↓ Size →	Capillary	0.158 cm 1/16 in.	0.317 cm 1/8 in.	0.635 cm 1/4 in.	0.952 cm 3/8 in.
Inside diameter, cm	.02-.05	0.119	0.165	0.393	0.800
Maximum Plates/m	3300	330	260	160	100
Max. practical length, m	92	18	18	18	31
Amt. liquid phase, wt %	--	3	5	10	20
Liquid film thickness (μ)	1	5	5	10	20
Particle diameter range (μm)	--	100/120	80/100	60/80	20/40
Permeability, x $10^{-7} cm^2$	200-800	1.5	2	3.5	4.0
Avg. linear velocity, cm/sec	25	10	10	7	7
Max. sample size, μl	0.010	1.0	2.0	20	1000

is certainly the case, for example, in most "matched columns" prepared for dual-channel programmed-temperature GC units.

The column packing material consists of a stationary liquid uniformly coated on a narrowly sieved solid support (usually a firebrick or diatomaceous earth). The material is then loosely packed into the column employing a variety of techniques such as vibration, vacuum, etc., so that no channeling or areas void of packing occur in the column. The ideal capillary, or (2) wall-coated column would be one in which the entire inner column wall is coated with a uniform thickness of the stationary phase. For more details on Column Preparation see Appendix F.

With little effort, GC columns can be maintained so that they will deliver reliable analyses. During use, the average column builds up contaminants from heavy samples that, after a time, can cause deteriorating column performance and possible ghost peaks on the recorder. This can often be corrected by "steam cleaning" (3). Steam cleaning can be accomplished by heating all of the GC sample system above 100°C (150°C usually preferred by authors) and injecting several 10-20 μl samples of water. If a flame ionization detector (FID) is used, water should be injected through the inlet until little or no organic material is desorbed from the column by the polar steam. This results in essentially no detectable FID response as the water passes through. Periodic weekly steam cleaning of a system will greatly reduce the need for extensive replacement of inlet systems and connecting lines and will usually increase useable column life.

Be sure to know the temperature limitations of the stationary phase being used (see Table 10-1) and keep all operations at least 10-15°C below this critical temperature. Observance of this criterion will certainly add to useful column life and prevent the detector and other parts of the unit from contamination with the "bleeding" stationary phase. A sure sign of excessive column temperature is not always gradual column bleed as evidenced by a slowly upward drifting baseline, but by large, broad and irregular shaped peaks. Typical large, irregular peaks are shown in Figure 10-1.

To keep the column free of foreign materials and moisture, a constant, dry, carrier-gas flow rate should be maintained. Another aid is to store the column sealed with a small positive pressure of the dry carrier gas. This can usually be accomplished by capping the ends of the column with suitable silicone rubber plugs after a constant gas flow has been maintained to purge the column. If possible, a column should be given a good steam cleaning and high-temperature purge before storage. This will generally eliminate foreign matter in the column that could cause

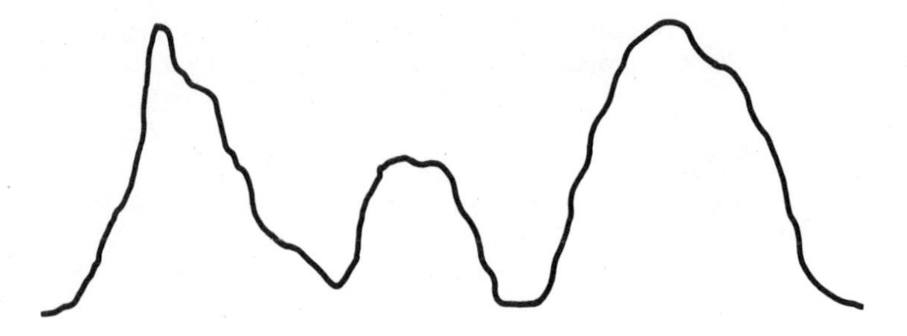

Note: Taken from actual analysis of gasoline on a 200-foot squalane capillary when temperature "mistakenly" reached 160°C.

Fig. 10-1. Typical column "blurps" (6).

degradation of the stationary phase during storage.

A drifting baseline can be caused by a new column not properly conditioned (i.e., still bleeding traces of stationary phase and solvent). To isolate the problem disconnect the column from the detector. If the drifting stops and the recorder pen returns to baseline, column irregularities are indeed causing the drift and can be eliminated (or improved) by using one, or all, of the following suggestions. If drift continues after disconnecting the column, the problems are electrical and will be discussed in subsequent chapters.

1. Recondition the column by disconnecting it from the detector, increasing the carrier gas flow and setting the temperature near the stationary phase limit. Continue to condition the column in this manner overnight. Then cool it to the original operating temperature and pressure, and reconnect to the detector. If baseline drift persists, proceed to suggestion #2.
2. The drift could be from column contamination or deterioration. Try steam cleaning as described previously in this chapter. If drift continues after this treatment, the column may be badly deteriorated or, for some reason, unstable, and should be repacked, recoated or replaced.

    Deteriorating resolution is usually caused by a deteriorating or contaminated column, or a leak in the system. Quickly leak check the system as described in Chapter 8, especially if it has been operated under widely varying programmed-temperature conditions. A good steam cleaning should be tried next. If resolution remains poor, the column should be replaced or recoated as required.

Fluctuating resolution is usually caused by an unstable condition within the column, such as channeling in packed columns or deterioration of the coating in capillaries. The solution to this problem is usually replacement of the column. This problem could also be caused by leaks, dirty connecting tubing and by fluctuating or cycling column temperature or pressure.

A decrease in retention time is usually caused by column deterioration and is solved by replacement or recoating of the column. However, a false low-temperature reading also could cause a loss in retention time.

Ghost peaks can come from areas other than the column, such as the septum or inlet system. However, ghost peaks can also originate in the analytical column when it is contaminated with heavy or polar materials. The best overall treatment for the appearance of ghost peaks is steam cleaning of the chromatographic system. If this doesn't solve the problem, column replacement or recoating is in order.

Peak height changes usually occur because of column changes that result in peak broadening or reduced column temperature. Thus, column condition, flow, and column temperature should be checked to solve this problem. Oxidation of the stationary phase or retention of water could also result in peak broadening.

Following are some general aspects concerning the maintenance of GC columns. To begin with there is no general treatment or single solution to chromatographic column problems. However, following are a series of suggestions which should greatly aid in the preventive maintenance of a gas chromatographic column.

1. Do not operate the column close to the maximum temperature which the stationary phase will stand. Many of the stationary phases will slowly decompose at these temperatures, and usually, columns will have to be replaced more frequently than is necessary.
2. Try to keep the column clean. Do not allow the column to become contaminated with heavy residue of samples that were analyzed. Use a backflushing arrangement, or alternately, use disposable pre-columns at the column head and change them with reasonable frequency when possible. This will prevent the heavy materials from trailing through the column and eventually

into the detector, giving contamination of the detector and a persistently poor baseline during operation of the column.

3. Gas chromatographic columns very often do not give the same performance when reinstalled in a GC unit as they did before they were removed. One common reason for this is oxidation of the stationary phase. When removing a column from the GC unit, it is preferable to cool it down to room temperature, disconnect the column while still full of carrier gas and quickly cap it at both ends so that it is sealed with an inert gas inside. In this condition, the column should be quite stable during several months of storage.

To conclude the discussion of columns, the configuration in which the column is coiled should be considered. The two most common coiling configurations for packed columns are shown in Figure 10-2. With the carrier gas continually flowing through the column, the force of gravity pulling downward on the column packing could result in channeling, with the observed effect being a noticeable loss of selectivity and resolution. Thus, the vertically coiled configuration (see Figure 10-2B) would be the best to use, as the packing could only channel at the top and bottom of the coils. The horizontally coiled column (see Figure 10-2A) could exhibit channeling throughout the length of the column. Because of the generally short length and adequate distance between coils of packed columns, uniform heating of the coils offers no difficulty. Often, however, the choice of vertical or horizontal coiling is made depending on the type of oven

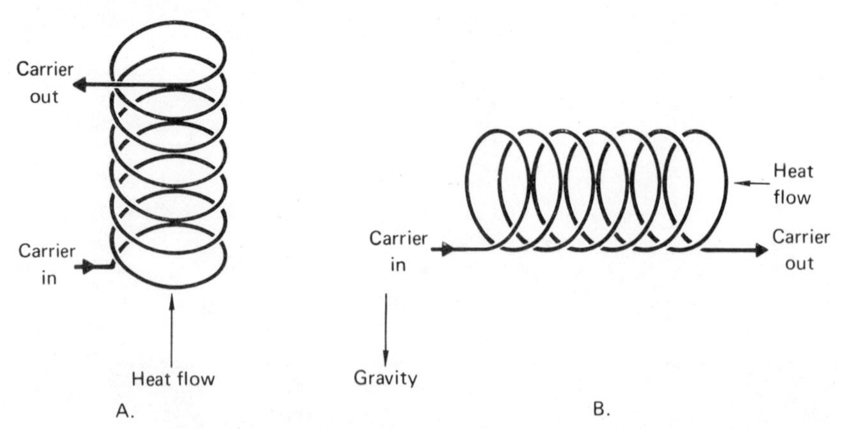

Fig. 10-2. Most common (A) and (B) packed column configurations (4).

design the columns are to be used with. Some column ovens require horizontal coiling so that the heated air will move about the column in a cycloidal fashion or will move axially through the column coil (7). Other ovens require the column to be coiled vertically to obtain the same thermal effects. A suggested minimum coil diameter to which columns should be wound has been previously described.

The configuration in which a capillary column is coiled will not result in inner disturbances such as channeling; however, the correct method of coiling capillaries is paramount in obtaining even heating. Capillary columns are usually wound on some sort of mandrel, from a common tin can to a spaced-mandrel type of coil. Winding a 100-foot capillary on a one piece mandrel such as a tin can is a neat and space-saving technique; however, with two or four layers of the column coiled on the can, uniform heating during temperature-programming is difficult to obtain. At a given isothermal temperature setting, the interior coils of the column will eventually approach thermal equilibrium with the outer coils by heat conduction and fairly even heating will result. However, if the temperature is to be programmed, there is little chance the inner coils and outer coils could register the same temperature as the heat input rises. To prevent such thermal gradients across the column, each coil should be separated by an air space to allow good circulation and obtain maximum benefit of column temperature programming. The ideal capillary coiling configuration is shown in Figure 10-3. Sometimes referred to as "basket-coiled" (8), such a configuration allows fairly even heating of each coil of the extremely long (up to 1000 feet in some cases) capillary columns. The preparation of a similar system has been described previously in detail (9). Such spaced-mandrel winding should ideally provide the best possible uniformity in heating, assuming a uniform oven heat flow. Some capillary chromatographers claim that loose coils of capillary tubing held together with wire can be temperature-programmed with good results. This may be true, assuming the coils are not too tightly wired together; however, there invariably exist some temperature gradients where the coils contact one another. In general, the column, whether packed or capillary, should be coiled to prevent contact with adjacent column coils and to enhance good air circulation around all external areas of the column.

A good column oven design should include adequate precautions to ensure that the oven compartment is not affected by other heaters, such as the inlet or detector heater. The column oven should be free from the influence of changing ambient temperatures and have a well-designed and adequate air flow system to maintain good temperature control of the

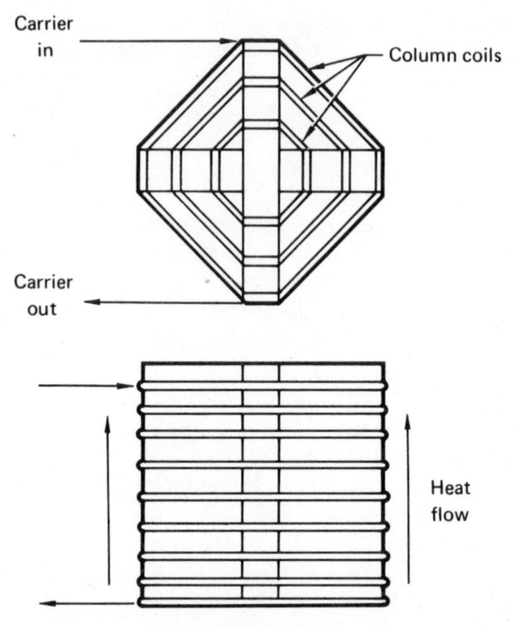

Fig. 10-3. Mandrel winding of capillary columns (8).

column. In the actual design of commercial ovens, many of the criteria stated above are met (11, 12, 13). The generally accepted way to heat a column to the desired temperature is by use of an air-circulating oven, although other techniques can be used such as: immersion in a liquid oil circulating constant-temperature bath; stagnant air heating (such as closed ovens with no fan, or covering the column with a simple heating mantel); or by direct passage of electrical current through the metal column walls. Most air circulating ovens are generally of the designs shown in Figure 10-4. In Figure 10-4A, the air is blown past the heating coils then through the baffles that make up the inner wall of the oven, past the column and back to the blower to be reheated and recirculated.

The air-flow configuration in Figure 10-4B is exactly opposite that of Figure 10-4A in that the air is sucked through the wall baffles, past the heating coils, and the freshly warmed air expelled by the blower. The blower drive motor should be well insulated from the oven environment, since not many electrical motors could withstand heat leaks with inside oven temperatures greater than 200°C. In general, rapid circulation of heated air within the column oven is essential. High-capacity, squirrel-cage blowers and carefully located baffles provide this, ensuring homogeneous temperature distribution, accurate thermal regulation, and

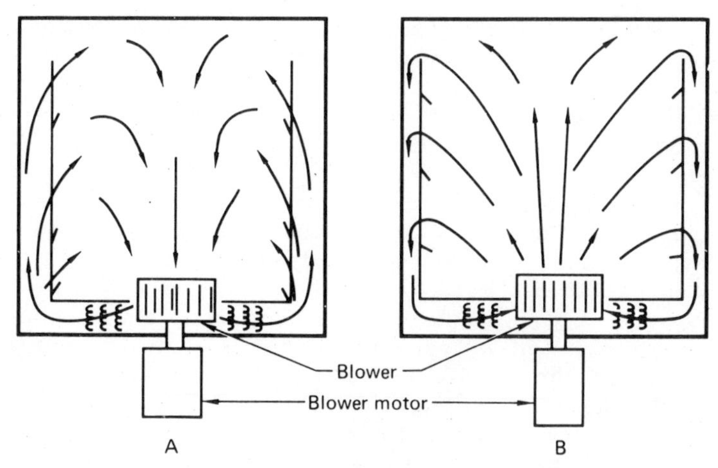

Fig. 10-4. General oven design (10).

rapid cooling when desired. Most column ovens are constructed of low-mass stainless steel to permit rapid heat-up and cool-down, and allow a uniform temperature to be maintained at ±1°C of the isothermal desired value and ±2°C of the desired temperature during programmed operation (10). There is usually little influence on the oven temperature from detector or inlet system heaters above 50°C, and programmed heating is usually quite linear. Another desirable feature of better ovens is that they are in separate modules from the electronic components. This is not to protect the column oven itself, but rather to protect the electronic components from heat that may escape from the column oven during high-temperature operation.

A GC column may be heated by passage of current directly through the metal walls of the column to produce the desired column temperature (11). For this type of operation the column is usually placed in a draft-free enclosure. However, this method of column heating requires electrical insulation from other chromatograph parts. Another method of directly heating the column is by wrapping it with heating tape or resistance wire; however, this is generally not desirable, since column cool-down takes an inordinately long time.

For essentially trouble-free operation, the column oven should be checked periodically, and minor maintenance performed. At least once every six months (especially under heavy usage), the blower motor (if not of the sealed bearing type), the oven lid, and other moving parts should be lubricated. To prevent sudden power failures, periodic checks of the power connection to the oven heater should be made.

Also, the actual heating wires should be periodically checked for signs of corrosion or deterioration, especially if the GC unit is situated in a somewhat corrosive atmosphere. Replacement of an obviously corroded or eroded heater wire may save time since a complete burn-out could occur during an analysis.

Another factor that could contribute to poor oven operation is a failure of the temperature-sensing thermocouple of the controller. Be sure that it is firmly installed in the proper place in commercial ovens. In homemade ovens, it is paramount that the thermocouple be exposed to the oven air in an area adjacent or near the column to obtain an accurate indication of the actual column temperature. Some workers attach the sensing thermocouple to the column itself, but this is not as accurate a technique for sensing total column temperatures, unless you can be sure the spot on the column to which the thermocouple is attached is indeed representative of the whole column. In addition, if operating consistently at high temperatures or in any type of corrosive atmosphere, the thermocouple wire should be watched closely for corrosion damage which could result in false temperature indications.

It should be beneficial to most chromatographers to consider a few oven system problems that have been encountered. For example, what is the procedure to follow when you set the column oven to heat to 150°C and nothing happens? First, check the obvious and make sure all electrical connections to the oven from the controller are intact. Make sure that no power cords have been accidentally pulled out or vibrated loose. If all of the connections are intact, and the thermocouple is reading the correct temperature (probably room temperature in this case), use a simple AC voltmeter to determine power is present at the temperature controller output. Also, check the oven heater for continuity to establish that the column oven does not have an open heating element. If all of these tests are negative, except that no power is coming from the controller, repair of the temperature controlling unit is required. If an open element is found in the oven heater wires, this is generally quite easily replaced in the laboratory with conventional spring-steel heating wire. For troubleshooting of silicone controlled rectifier circuits, see Appendix D.

Fluctuating-column-temperature can be a much more complex problem. One of the most likely causes of this phenomenon could be an intermittent leak to ambient air where room air could cause changes in oven temperature. A faulty or corroded temperature sensing thermocouple could cause temperature fluctuations, especially if programming the temperature might cause distortion of the thermocouple itself.

Generally, a good cleaning and resoldering of the thermocouple tip will take care of this problem. Another possible cause of fluctuating column temperature could be trouble in the oven temperature controller. Probable solutions to this difficulty will be discussed in the next section. Cycling or fluctuation in column oven temperature can also be caused by fluctuating line voltages.

Overheating has ruined many analytical columns. To prevent this, an overheat protection system is necessary. The most common is thermal fuses made of alloys that melt at certain predetermined temperatures, completely shutting off power to the oven. Most modern controllers and programmers have temperature limit shut-off switches which consist of a microswitch attached to the pyrometer. When the sensed oven temperature reaches this pre-set level (selecting by setting the maximum temperature set point on the pyrometer), the microswitch opens, cutting power to the oven. Such a system, operating from the measured column temperature is superior to any system operating through the programmer or controller readout temperature. Usually this upper limiting switch has an accuracy of $\pm 10\%$. Several different types of overshoot mechanisms are employed with new solid-state programmers.

Temperature overshoot when approaching an isothermal point of operation is usually caused by poor air circulation in the oven, especially if the temperature required is considerably above that at which the column had been operating. This overshooting effect is much more pronounced with on/off full-power type controllers than the power proportioning type. If a stagnant air oven (such as a heating mantel) is being used, temperature overshoot when changing to a higher column operating temperature can be minimized by a low power setting on the controller, or if a programmer is being used, slowly programming to the desired temperature.

Temperature overshooting is also caused by a stalled oven blower. This is usually attributable to failure of the blower motor, caused by excessive heat, worn bearings, etc. Use a voltmeter to determine whether or not the motor is receiving power; if not, check all connections and make necessary repairs. This is easily done from schematics supplied with most commercial units. If all these tests show no malfunctions, the motor must be replaced or repaired; and, except for lower temperature, isothermal operation, the chromatograph is "down for repairs." Removal of most blower motors is usually quite easy after the blower has been removed from the shaft.

Poor oven lid seals would result in temperature cycling and inability to hold high temperatures. The obvious solution to this problem is replacement of the faulty seals.

One of the most common problems with column ovens is

that of not being well enough insulated from other heated components of the chromatographic system (vaporizer, detector, splitter, etc.). This is especially true if the column must be operated at 50°C or less, as these components usually must be operated at least at 200°C or higher for good performance. Even with the presence of large thermal barriers some heat is always passed by conduction through the connections to the column. Thus, a completely isolated, yet usable, oven is hard to achieve.

All of these different column heating devices should be interfaced with an accurate and sensitive temperature controller. The controller should sense column oven air temperature continually and supply the amount of power required by the oven heating coils to obtain the desired heat input. In general, three types of temperature controllers are used: the constant voltage unit, the on/off full power unit, or the power proportioning unit. The most common type of constant voltage controller is the conventional powerstat. In early GC units, the column temperature was often crudely controlled by a powerstat; however, without proper temperature sensing and feedback devices, the column temperature could not be well controlled. The earliest feedback type of unit was the full-power, on/off type of controller. A temperature setting was selected and the temperature measured accurately by various types of sensors. The sensor would supply feedback information to the power supply for the oven heater, either calling for heat or not. Thus, the oven would be kept at a given temperature by cycling the heater on and off at full power as required by the sensor. However, with this type of control, small cycles in temperature sometimes developed resulting in inconsistent column operation. The constant voltage type of controller supplies a constant voltage at a reduced level to provide the desired temperature (i.e., reduced from full line voltage). The unit is isolated from fluctuations in line voltage by an isolation transformer. Such controllers are usually quite accurate if adequate insulation of the column from ambient air changes is provided. Power proportioning controllers are generally more sophisticated forms of constant voltage units, employing solid state circuitry and rapid time-constant sensing circuits.

If the temperature of the column oven is to be programmed, the programmer should supply increasing power to the heating coils instantaneously on command of the controlling unit. There are also several types of power and control units for programming the column oven temperature, from simple mechanical drive units that supply heat required by cycling full oven power, to advanced power proportioning models, where heating rates are generated electronically by complicated solid state circuitry. The more popular and less expensive mechanical drive programmers have rates selected by various gear ratios, the gears being driven by a constant-speed synchronous motor. After calibration procedures, so that the temperature read-out for the digital mechanical system is consistent with the measuring thermocouple or sensor, a feedback loop allows supply of the amount of power required to establish the programming rate. Electronic programmers depend mainly on different solid state time-logic RC circuits to increase the power to the column oven to supply the desired heating rates.

Temperature controllers and programmers are relatively easy to maintain, especially if they are of the new solid-state variety, where operational heat, such as found in tube-type units, is absent. In general, a calibration check monthly or as prescribed by the manufacturer is sufficient to determine efficiency of operation. Calibration procedures are all defined in operations manuals for various types of programmers and controllers, and will not be discussed here. However, one quick way to check the accuracy of the sensing thermocouple circuit is to immerse it in boiling water and adjust the meter to 212°F (100°C). If the unit is fitted with a subambient unit, check the thermocouple in ice (32°F or 0°C), and inspect the coolant relay operation. If the unit has a mechanical drive for the temperature program (servo-motor plus a gear system), periodically check the gears for signs of wear or missing teeth. In a mechanical drive unit, never force-mesh the gears together while the servo-motor is on, as this will generally result in gear damage. A little care when selecting programming rates in gear driven units will go a long way toward prolonging the life of most programmer gear systems.

Temperature controllers and programmers result in a good portion of the problems encountered in the column oven system. Temperature fluctuations are frequently caused by faulty or failing controllers or programmers. If the rest of the oven unit checks out as described previously in this section, then the cause for this problem obviously lies in the controller. Some suggestions might be:

1. Relocate the thermocouple in the oven to an area of greater exposure.
2. Recalibrate the unit, as synchronization between the required temperature and actual temperature is necessary.
3. Thermocouple circuit may be picking up electrical "hash." This is evidenced by vibration of the pyrometer needle. Re-route or shield the thermocouple leads.
4. Check reference voltages to and from the sensing

units as described in most manuals (especially the output of SCR power-proportioning systems).

5. The sensing unit could be inadequately damped, and the problem could be cured by additional damping.

If the oven will not heat, the trouble could lie in the controller. First of all, check to see that the temperature setting is high enough to bring power on. Check the thermocouple circuit. Finally, check the output voltage of the controller as described in the user's manual.

Overshooting the selected temperature is often caused by poor calibration or by the unit being overly damped, thus, not sensing the temperature changes rapidly enough to control at the selected temperature. Solve this problem by decreasing the damping.

When programming the temperature, a common malfunction observed is stepwise jumps in temperature. This is usually caused by one of two things: if mechanical drive, the gears are slipping, or the unit is either not calibrated correctly or is over-damped. These are all easy items to rectify.

A nonlinear program rate usually suggests that the servo-motor for the mechanical drive unit is malfunctioning, or that there is an intermittent leak in the air bath thermostat. In power proportioning units, electrical malfunctions within the programmer are possible. This usually is indicative of faulty voltages being sensed by the controller units, and requires expert attention.

One last item to consider in the column oven system is sample transfer lines between the vaporizer and column and from the column exit to the detector. For optimum operation of any GC system, this connecting tubing should have as small a volume as is practicable due to the fact that all connecting lines constitute dead volume. In most commercial and homemade units these connecting lines are usually too large, too long, and especially too cool. Only 1/4 inch of unheated 1/8 inch o.d. stainless steel tubing can result in a 25°C temperature drop. Transfer lines larger than 1/8 inch o.d. should never be used. Instead, well-heated, large bore capillary tubing is preferred. Good GC design demands short and efficient sample transfer lines.

Routine maintenance of sample transfer lines is equally as important as the other areas of the column system. When the system is cool, and with installation of each new column, all connections should be checked for gas leaks with soap solution. Steam cleaning of the column as described previously should also do an adequate job of removing adsorbed materials from these lines. If very heavy or corrosive samples have been run, such as pesticides or alkyl-lead compounds, remove these lines periodically and flush with solvents. If large amounts of deposits are on the walls (determined by running a rigid wire through the tube), replacement is recommended to prevent selective removal or retardation of polar compounds. Check heater operation in these areas to make sure that good thermocouple contact and placement are made.

Problems in the sample transfer lines such as deposits, leaks, and cold spots can result in many of the symptoms observed for inlet system and column malfunctioning (i.e., less resolution, loss of sample, peak broadening, peak shifting, peak disappearance and ghost peaks). Leaks, of course, can be detected rapidly with soap solution and stopped; however, if the systematic elimination of trouble spots points to the sample transfer lines, they may be steam cleaned or removed and solvent cleaned. But, in light of the time involved, replacement with new, clean tubing is usually the least time consuming and expensive.

In review, accurate troubleshooting in the column oven system can save operating time. A systematic approach, using any combination of the hints and experiences listed in this chapter and in the chapter on Comprehensive Troubleshooting should help reduce down time, and allow the chromatographer to become more familiar with his particular GC unit.

REFERENCES

1. J. Q. Walker, *Hyrocarbon Processing, 50*, No. 10, 99 (1971).

2. L. S. Ettre, *Open Tubular Columns in Gas Chromatography*, 1st ed., Plenum Press, New York, 1965 (pp. 79-82).

3. Research Notes, published by Wilkens Instrument and Research (Now Varian-Aerograph), Fall Issue, 1961.

4. Private Communication, J. B. Maynard, 1971.

5. No. GC-AP-010, P10 Literature on Oven Designs for Gas Chromatography, 1967, Perkin-Elmer Corporation, Norwalk, Conn.

6. Private Communication, J. B. Maynard, 1971.

7. J. C. Giddings, *Anal. Chem., 36*, 1580 (1964).

8. D. J. McEwen, *Anal. Chem., 35,* 1636 (1963).

9. J. Q. Walker, *Hydrocarbon Processing, 49,* No. 6, 178 (1970).

10. Private Communication, J. Q. Walker, 1970.

11. Perkin-Elmer Model 810 Operation Manual, Perkin-Elmer Corporation, Norwalk, Conn.

12. Hewlett-Packard Column Oven, 7620, Operation Manual, 1970, Hewlett-Packard Corporation, Avondale, Pennsylvania.

13. Varian-Aerograph, Previews and Reviews, No. 5, Walnut Creek, California, 1967.

14. Private Communication, J. B. Maynard, 1971.

15. W. E. Harris and H. W. Habgood, *Programmed Temperature Gas Chromatography*, Wiley, New York, 1966.

QUESTIONS AND ANSWERS

Question: Is it generally necessary to silylate glass columns before packing them for use in analyzing generally unstable materials?

Answer: Probably not. Whether or not a glass column should be silylated depends on the type of samples to be run. For example, if the sample to be analyzed contains material that is highly polar, has a high molecular weight and may be somewhat unstable if encountering any active sites when passing through the column, a silylation step may be in order.

Question: Early eluting peaks are well-resolved during a GC run, but later peaks do not appear at all, or they are very broad and diffuse. What can cause a problem such as this?

Answer: The most probable cause is that the column temperature is too low and/or the column pressure is too low; however, the effect of column temperature would be much greater than the pressure. On the other hand, there could be some adsorption on active sites in the column (which might be solved by increasing the column temperature), or there could be some chemical reaction with the stationary phase causing essential loss of the heavy components. If this is the case, silylation of the column and inlet system might be to some advantage.

Question: What are the main advantages of preparing derivatives before GC analyses?

Answer: Derivatives are prepared for certain classes of compounds to make them chromatograph better. For example, carboxylic acids are extremely polar and hard to determine quantitatively by GC, even on polar stationary phases. Thus, by very simple and rapid procedures, the methyl esters or trimethylsilyl derivatives are prepared to make the acids much less polar, and to increase their effective volatility. Then, in many cases, the derivatized materials can be chromatographed on a boiling-point column such as any of the high-boiling silicone-based stationary phases. Only by derivative preparation can many high-boiling, polar and unstable compounds (such as many materials of interest in biological laboratories) be chromatographed at all. The preparation of derivatives should always be considered when faced with chromatographing polar materials such as carboxylic acids (as methyl esters and trimethylsilyl ethers). Other derivatives can be prepared that not only considerably lower the polarity of the material, but, enhance its sensitivity to certain specific detectors, such as the electron capture detector (ECD). For example, to effectively determine very low (ppm) amounts of heavy amines, the amide derivatives of trifluoroacetic anhydride (TFA), pentafluoropropionic anhydride or heptafluorobutyric anhydride can quickly be prepared and analyzed at very high sensitivity using an ECD, as these fluorinated derivatives are quite electronegative. The TFA esters of alcohols can also be prepared to improve their behavior on a GC column. In general, derivatives should be considered when the resultant derivative would render a material (1) more volatile, (2) less polar, (3) more stable and (4) more easily detected in trace amounts with specific detectors.

Question: What is the best GC technique to determine low concentrations of water in organic materials?

Answer: The best technique from the standpoint of accuracy, short analysis time and reliability is analyses on a 6-foot x 1/8-inch O.D. stainless steel column packed with Porapak Q, 80/100 mesh size. The water elutes before propane, and thus, the water content of even gaseous industrial streams can be fitted with a backflush valve; and, as soon as the water peak is measured, the heavy organic material can be backflushed off the column. This is a very powerful qualitative and quantitative technique for determining water in organic materials.

Question: In trying to chromatograph nanogram amounts of mercaptans and hydrogen sulfide, it was found these small amounts of materials are almost always adsorbed in the column. Is there any other way to do such an analysis other than deactivating any active sites with TMS and/or using a Teflon column?

Answer: No, not of any proven significance. Of major importance is keeping the column as free of polar sites as possible; and, depending on the individual case, Teflon or derivative-deactivation may be the only way that very small amounts of these materials can be chromatographed without adsorption. However, one good column for the separation of light sulfur compounds is Porapak Q-S.

Question: Using an F and M Model 810 GC unit, three times there have been oven temperature runaways in excess of 450°C during programmed-temperature operation, destroying the column in the oven by essentially stripping stationary phase from it thermally. After cool-down, everything seemed normal. What could have caused the temperature runaway?

Answer: The F and M 810 GC unit probably has a mechanical drive temperature programmer that advances upward at the temperature rise-rate selected until the temperature set-point (or pyrometer readout) reaches the preset upper limit. At this point, a microswitch is closed when the set-point and upper-limit switch engage, either holding at the upper temperature limit or turning off the column heat, and effecting column cool-down. The most common problem that causes temperature runaways in programmed-temperature operation is corrosion of the microswitch, and the programmer is never told to stop its upward travel. However, if this is isothermal operation, the temperature controller is continually calling for heat. This could result from (1) a faulty thermocouple, (2) a faulty temperature platinum sensor element or (3) a bad silicone-controlled-rectifier (SCR) that is not properly proportioning power, but is continually supplying full power to the heaters.

Question: With a Hewlett-Packard 5750 GC unit, the fan motor failed to operate after being shut off for a column change. The fan fuses were not blown; there was current across the fan-motor switch, and the squirrel-cage blower could be rotated freely by hand, demonstrating no mechanical binding. The service engineer was called, and the fan operated properly the first time it was turned on. However, there were subsequent failures when the column oven was cooled down completely again for another column change. What could be causing this intermittent loss of power to the fan motor?

Answer: This is most likely a case of intermittent electrical contact, especially if the oven has been operated at high temperatures. There are ceramic insulators under the column compartment that keep the blower electrical power leads from becoming too thermally hot. Sometimes these insulators crack, and with wide temperature swings, such as cool-down periods, sometimes short to ground or lose contact completely, causing the fan motor to lose power. After the temperature equilibrates and/or the contacts are possibly jarred into touching again, the oven fan motor will once again have power, and will run. We have experienced a similar failure with the fan motor in a Hewlett-Packard 7620 unit caused by these cracked ceramic insulators.

Question: How do you precisely determine the weight percent loading of a packed GC column?

Answer: Weigh out a precise amount of stationary phase, then dissolve in a suitable solvent. Weigh out the desired quantity of solid support to give the desired loading, and mix with the stationary phase-solvent mixture. The actual weight percent loading can then be obtained by reweighing the solid support impregnated with the stationary phase, or trapping the stationary phase stripped off along with the solvent. The weight of the stationary phase impregnated on the solid support can be obtained by difference, and the weight percent calculated.

Question: What would be a good column system to use to analyze for hydrogen and CO_2 in the head-space vapors in a can in only one analysis? At present two analyses are required, one to determine the hydrogen and one to determine CO_2.

Answer: There are several possible ways to do this. The easiest would be to use a double-pass type of analyses through a thermal conductivity detector (TCD). Nitrogen must be used as the carrier gas to get a suitable TCD response for hydrogen. The analysis could proceed as follows: The air peak (H_2 + N_2 + O_2) could be separated from CO_2 on a Porapak Q column, with the effluent passing through the positive (sample) side of the TCD, resulting in two peaks. The effluent from the detector could be directed to a 5A molecular sieve column where the H_2, N_2 and O_2 would be separated, and the CO_2 adsorbed. This effluent could be directed to the other side of the TCD to determine the H_2 content. Since N_2 is being used as the carrier gas, a negative peak will be obtained from the H_2, but since the H_2 is being detected on the negative (or reference) side of the detector, it will appear as a positive peak on the recorder. Suitable calibration with standard gases should allow quantitative determination of the H_2. Nitrogen and most of the oxygen would not be observed, as nitrogen is used as the carrier gas. Following is a drawing of what such a chromatogram could resemble.

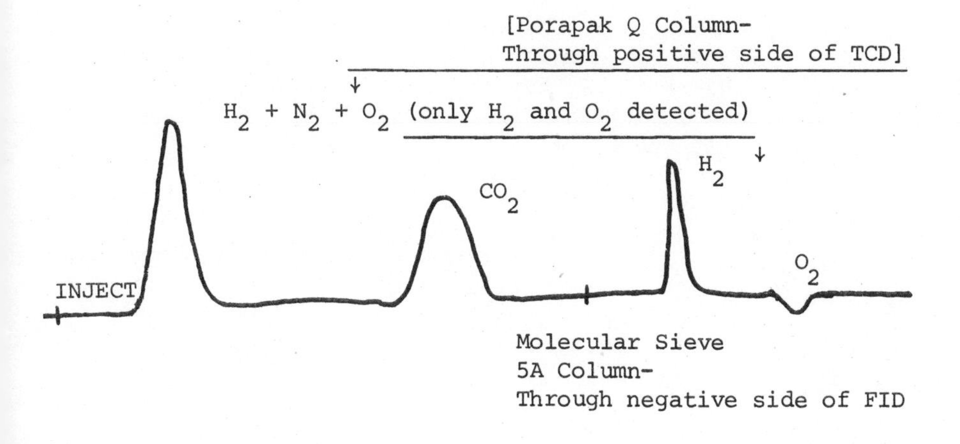

Another, more complicated approach would be to use a switching valve after the Porapak Q column to divert the hydrogen-containing peak to the molecular sieve column, and the elution of the H_2 and CO_2 suitably timed to show no peak overlap. The main problem here would be measuring the negative peak that would be obtained for hydrogen.

Question: What is the most practical means for cleaning a column of impurities having an electron capture detector (ECD) that has been repeatedly used for analysis of chlorine containing components?

Answer: Probably the best way to clean such a column is by repeated injections of water (steam cleaning); however, the column *must* be disconnected from the ECD. The ECD could be cleaned with the solvents suggested by the manufacturer. A dirty column situation could generally be prevented by a precolumn (6-foot x 1/8-inch O.D.) packed with a material that would pass the chlorine-containing component and hold up any heavy material that could contaminate the column. This precolumn should be changed often to prevent any dribble-over of heavy contaminants into the analytical column. Of course, this will also help to keep the detector clean.

Question: What is the maximum length of tubing that can be used between columns and backflush or switching valves without loss of resolution?

Answer: The key in this situation is to keep the dead-volume to an absolute minimum; thus, 0.02 or 0.03-inch I.D. tubing should be used. In most cases, 0.01-inch I.D. tubing should be avoided, as it can act as a flow restrictor on these low-pressure flows usually found in the vicinity of such valves. Another must is to keep the temperature of the connecting tubing at least 10°C above the boiling point of the highest boiling component to be analyzed. If 0.02 or 0.03-inch I.D. tubing is used, as much as 2-3 feet can be used as connective tubing with all column systems except 0.01-inch I.D. capillary systems, with no loss in resolution.

Question: What is the best way to couple GC columns in series with minimal loss of resolution?

Answer: One of the best methods is to use drilled-out unions, such that the column ends almost are butted together. For capillary columns, low dead-volume unions of 0.01-inch I.D. are available commercially and should be used whenever connecting capillary columns in series.

Question: Is Dexsil 300 a good stationary phase for the separation of polynuclear aromatic hydrocarbons, such as benz-a-pyrene and benz-e-pyrene?

Answer: The Phenyl-Dexsil 300 is best for separating polynuclear aromatic hydrocarbons. However, some initial concentration step, such as liquid-solid chromatography, is needed to produce a concentrated heavy aromatic cut for analysis on a Phenyl-Dexsil 300 column. This phase has the good high-temperature stability and aromatic selectivity needed for such an analysis.

Question: Is temperature lag a very big problem in programmed-temperature GC analysis?

Answer: Modern solid-state power-proportioning temperature programmers result in very little column oven temperature lag. The important thing is to prevent *column temperature* lag, and this is usually best done by making sure there is good air circulation around the individual column coils. This is especially important for capillary columns which are relatively high mass, especially when coiled tightly. Refer to Chapter 8 for suggestions in correct methods and techniques of column coiling.

Question: What could cause humped peaks such as shown in the following figure if the component being analyzed is known to be a pure component?

Answer: This could be the result of ghost peaks, as discussed in Chapter 14, or could be due to a sluggish recorder, syringe contamination or some component decomposition on the column. The most common cause of such peak is ghosting or sample decomposition on the column. Septum decomposition, to supply the ghost peaks, should definitely be considered.

Question: What would be the best way to prevent degradation of sensitive materials during a GC analysis; on-column injection or a glass column instead of stainless steel?

Answer: Both of the techniques mentioned in the question could be a factor in preserving the integrity of a sensitive compound during sample injection. Exposure to the high temperature in the vaporizer necessary to vaporize large, unstable molecules is a quick way to insure some kind of sample decomposition. In this case, on-column injection is definitely required; in fact, on-column injection should be considered when analyzing any kind of even slightly thermally unstable materials. The use of glass columns will definitely reduce the chance for sample decomposition while passing through the column when coupled with the right type of stationary phase and column temperature. Of these two factors, the use of on-column injection would probably be most important in insuring sample integrity on the column and throughout the analysis.

Question: How do you achieve "textbook" straight baselines when doing programmed-temperature analyses?

Answer: Perfectly straight baselines are always difficult and sometimes quite impractical to achieve. A drifting baseline most of the time has essentially no bad effect on the accuracy or integrity of an analysis, especially with the integration and computer systems available today that do such excellent jobs of compensating and correcting for drifting baselines. The most painless way to achieve level baselines throughout a GC analysis is to use the matched dual-column technique. This can be used with either thermal conductivity (TCD) or flame detectors. If the column temperature is to be programmed, good flow controllers are necessary so that column flow-rates do not change appreciably during an analysis, thus causing baseline drifting, which, of course, is more pronounced with a TCD than a flame detector. Another point to bring out here is to make sure the septa used have been extracted and vacuum-heat treated so that heavy materials cannot be leached from the septum to be eluted during high-temperature portions of the column temperature program (Table 8-1). Obtaining a flat baseline with a single column is a much more difficult situation. Here, usually a combination of temperature and pressure programming is used to insure the sharp elution of heavier materials instead of temperature programming only. With a single column, the temperature is programmed up to some level where significant column bleed has not yet begun, and then the column pressure is programmed manually or via a pressure programmer as required for the analysis. This technique can only be successfully applied with a flame detector with added carrier gas, as a TCD and a flame detector without added carrier gas are too flow sensitive to give flat baselines and constant sensitivity. In fact, this technique works best with a capillary column and a flame detector, as capillaries have small changes in flow with pressure changes; and, in addition, compensating dual-capillary columns are impractical due to the large variabilities present in preparing such columns. Once again, as with dual-column techniques, a nonbleeding septum is a must.

Question: How do you minimize the holdup of high-boiling impurities from previous injections of wide-boiling mixtures?

Answer: The best way is to raise the column temperature and backflush until a level baseline is obtained. If a backflushing facility is not available, a column bake-out at increased pressure and a column temperature slightly under the maximum for the stationary phase may do the job. This is not as effective or as quick as backflushing, but may suffice.

Question: If two compounds are analyzed on a variety of GC units with a variety of flame detectors, will the relative response of two compounds to each other be constant when moving from one instrument to another? Will individual compounds have the same response factors on different flame detectors?

Answer: Individual compounds will generally have different response factors on different flame detectors. Very seldom will two flame detectors have the same response factors for the same compound. Also, the ratio of response factors of two components will very seldom be the same. When quantitative data are required, to be on a firm quantitative basis, response factors should be determined when analyses are done on different flame-detector equipped GC units.

Chapter 11

GAS CHROMATOGRAPHY DETECTORS

As a sample component emerges from the column, the time at which it is eluted and the quantity being eluted are evidenced by the detector. The detector then relays electrical signals, via an amplifier, to the strip chart recorder establishing a permanent record of the analysis. The arrangement of the detector in the GC system is shown in Figure 11-1. Even though the primary function of the GC instrument is separation, the detector provides the vital link between the column separation and the chromatogram. It is therefore important that the chromatographer be well informed in the theory, operation, maintenance, and repair of detectors.

There have been more than twenty different types of detectors developed for use in GC. However, only a relatively few of these detectors are in common use today. Most of these detectors and their uses will be mentioned here, but only the more commonly used detectors will be dealt with in detail.

TERMS

No two detector types are operated identically, nor are they all applicable to every type of detection problem. However, some common terms are used to describe detector performance, and these can serve as a guide to select the proper

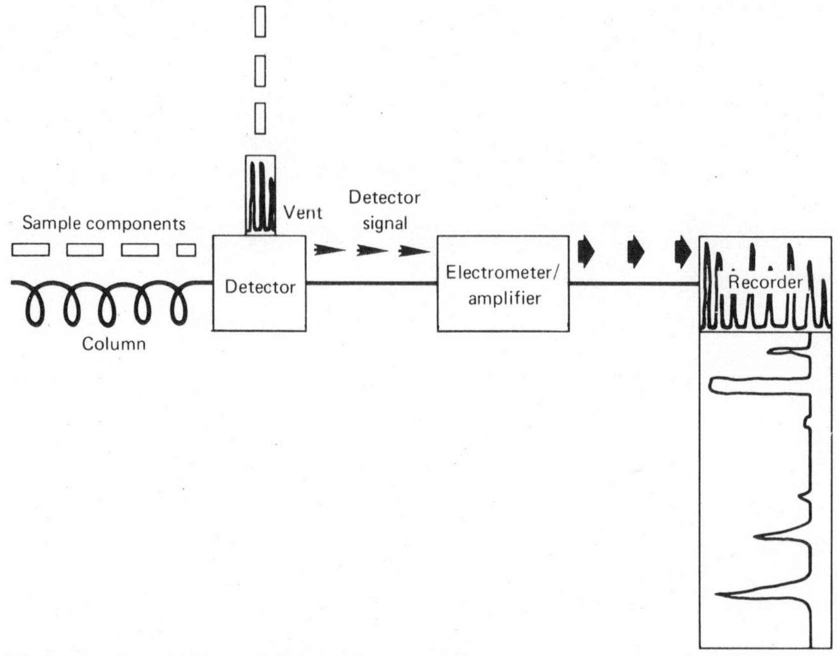

Fig. 11-1. This simple diagram shows the position of the detector in the GC system (1).

detector and evaluate detector performance.

Detectors are currently classified as belonging to one of two categories according to how they measure the sample. Detectors which respond independently of the flow rate of the carrier gas are called *mass flow rate detectors*. While, in theory, these detectors are not affected by changes in carrier gas flow rate, changes greater than about 25% will be significant. The carrier gas flow rate must be held constant for *concentration dependent detectors*. With these detectors, the greater the amount of carrier gas eluted with the sample, the less the response from the detector.

Detectors can be classified in terms of their *selectivity*. Some detectors will respond to almost any type of sample and are called universal detectors. Most detectors, however, will not respond to the presence of certain groups of compounds. In fact, some detectors are designed to respond specifically to a small group of compounds such as phosphates or halogens. The selectivity of a detector is illustrated in Figure 11-2. Column effluent from a sample containing air, carbon dioxide, carbonyl sulfide, hydrogen sulfide, propane, and propylene is split 1:1. Half of the effluent enters an electron capture while the other half enters a flame ionization detector (both are discussed in

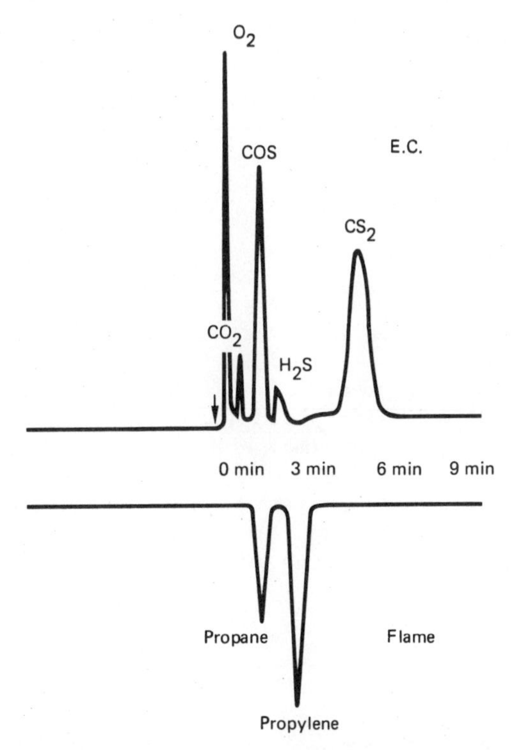

Fig. 11-2. The selectivity of detectors is illustrated by these two chromatograms showing the response of two different detectors to the same mixture. (Courtesy of Hydrocarbon Processing) (2).

detail later). Chromatograms obtained from these two detectors are vastly different. The flame ionization detector registered only the hydrocarbons, propane and propylene. The electron capture detector did not respond to the hydrocarbons at all, however, it did indicate the presence of the other sample components. Some detectors are highly selective and a are usually directed toward a particular purpose. Phosphorous detectors can be used, for example, to detect pesticides contained in the complicated mixtures resulting from the analysis of fruits.

Though the detector does not respond to a class of compounds this does not mean that the compounds will not have some influence on detector performance. For example, while flame ionization detectors do not respond appreciably to water molecules, the presence of large quantities of water in the sample can result in distorted peaks with some columns and lead to poor quantitation.

A quantity used commonly to describe detector performance is *sensitivity*. Essentially, sensitivity is a measure of the ability of a detector to signal the recorder of the presence of an emerging sample component.

Sensitivity measurements depend on the type of detector. The sensitivity of mass flow rate detectors is the peak area divided by the weight of the sample component as follows:

$$S_M = AS_RC/W$$

where,

S_M = Sensitivity of mass flow rate detector
A = Peak area (cm^2)
S_R = Sensitivity of recorder (m^V/cm)
C = Reciprocal of chart speed (min./cm)
W - Weight of component (mg)

The sensitivity of concentration dependent detectors must account for the carrier gas flow rate. Sensitivity for this type of detector is given as follows:

$$S_C = \frac{AS_RC/\mu}{W}$$

where,

S_C = Sensitivity of a concentration dependent detector ($m^V./mg$)
A = Peak area (cm)
S_R = Recorder sensitivity (m^V/cm)
C = Reciprocal of chart speed (min/CM)
$/\mu$ = Carrier gas flow rate at column outlet (cm^3/min)
W = Weight of component (mg)

Sensitivity is a convenient means of comparing detectors of the same type. It is difficult, however, to compare the sensitivity of a concentration dependent detector. When the peaks of a compound are used to measure sensitivity this is also significant and when comparing detector, the same column and peak should be used to determine sensitivity. Obviously, the electrometer/amplifier range and attenuation should be held constant also since this will directly influence the peak size.

The signal from the detectors can be amplified by nearly any amount. At high amplifications *noise* will develop which is characterized by rapid fluctuation of the recorder pen resulting in a chromatogram of the type shown in Figure 11-3. The noise (p) is measured over the maximum deflection of the recorder pen when no sample is eluting. Since this noise will make it difficult to measure small peaks accurately,

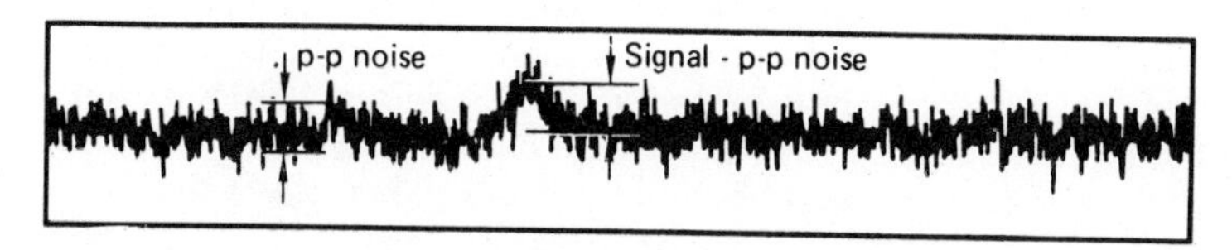

Fig. 11-3. The minimum detectable quantity is defined as a peak which is twice as high as the noise level. (Courtesy of Keithley Instruments, Inc.) (3).

the *minimum detectable quantity* has been defined as that quantity of sample producing a peak with a height of 2p.

Linear Range is important to accurate quantitative determinations. If a detector responds to a given amount of sample, twice as much sample should produce twice the detector response. All detectors have a range of concentration over which this relationship will hold true. To measure the linearity of a detector, the detector sensitivity is measured over a wide range of sample concentrations. The smallest concentration should be the minimum detectable quantity. If a semi-log plot is made of sensitivity as a function of the log of concentration, a line will result which is straight over the linear range. The upper limit of linearity is the point at which the curve deviates from a line. The *linearity* of the detector is then the upper limit of linearity divided by the minimum detectable quantity. Because a detector does not exhibit a wide linear range, this does not mean that the detector is of little value. It simply means that more calibration of peak area versus concentration will be required for accurate quantitation.

KINDS OF DETECTORS

The detectors used in GC must now be considered separately. Some have different application, operate on different principles, and require different precautions to be taken in their use.

THERMAL CONDUCTIVITY DETECTOR

One of the oldest and still widely used type of detector is the thermal conductivity detector (TCD) or katharometer. A gas, flowing across a heated filament, will cool that filament to some extent by absorbing some of the heat. Different gases have different abilities to cool the filament. If the current heating the filament remains constant, and, if a constant flow rate is maintained, the filament will soon establish an equilibrium temperature. This temperature can, in turn, be measured from the resistance across the filament. Should the composition of the gas flowing across the filament be changed, the ability of the gas to conduct heat away from the filament will be altered. The change in the temperature of the filament can again be measured by the filament resistance. This is the basic principle behind TC detectors.

In Figure 11-4, a simple diagram of a TCD cell is shown.

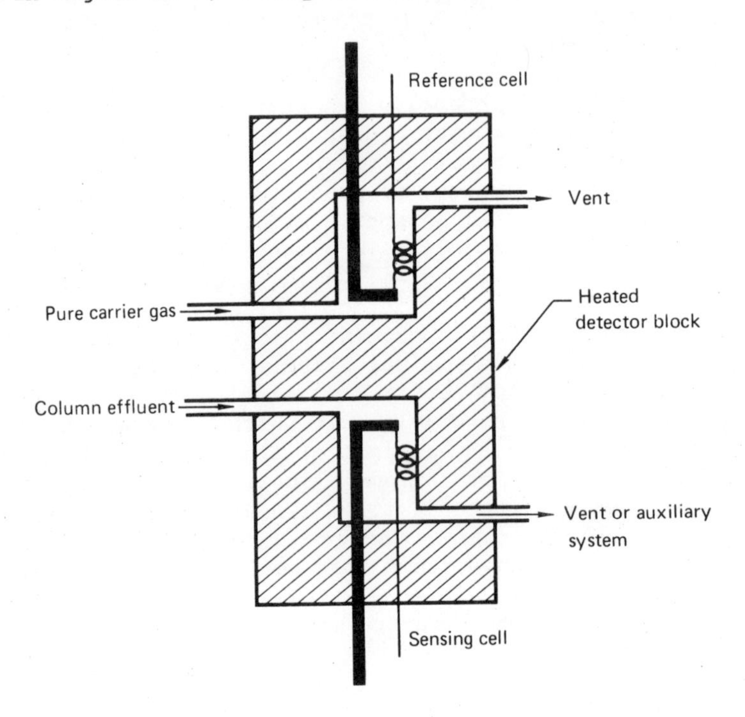

Fig. 11-4. A simple two cell thermal conductivity detector (TCD) (4).

The filament is contained in a heated block through which the gas is flowing. The filaments from the two cells are connected to form the elements of a Wheatstone Bridge, as shown in Figure 11-5. Through one cell, the reference cell, only carrier gas is flowing. Column effluent flows through the second cell. When no sample is eluting from the column, the temperature, and thus resistance (20-40 Ohms) of each filament will be the same, balancing the bridge. As sample, mixed with carrier gas, enters the second cell, the filament temperature will change resulting in an imbalance in the bridge circuit. The imbalance will, in turn, generate a signal which can be amplified and relayed to the recorder.

In practice, four filaments are used, two reference

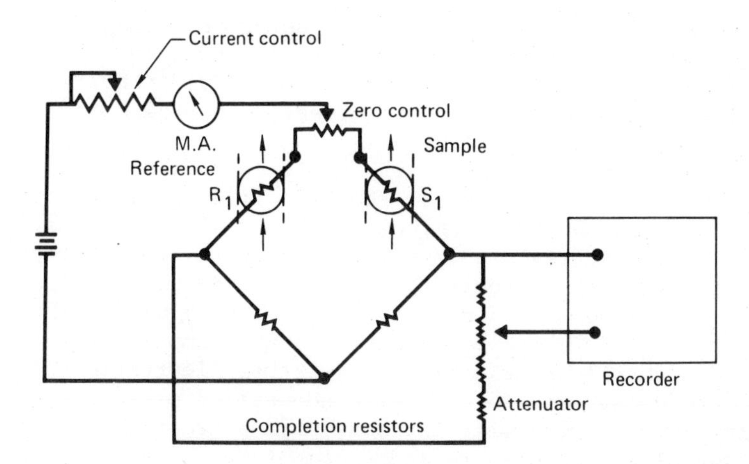

Fig. 11-5. The Wheatstone Bridge circuit for a two-cell TCD (Courtesy of J. of Gas Chromatog.*) (5).*

filaments and two sample filaments. These are arranged as seen in Figure 11-6. The advantage of this arrangement is that it provides twice the output signal and tends to stabilize the Wheatstone Bridge against outside temperature fluctuations.

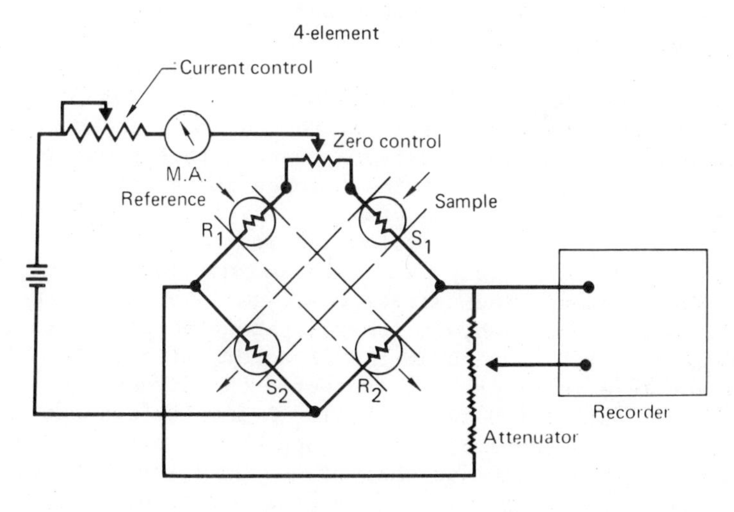

Fig. 11-6. In practice, most modern TC detectors have a pair of reference cells and a pair of sensing cells. (Courtesy of J. of Gas Chromatog.*) (5).*

Some consideration should be given as to the types of carrier gas to be used with TC detectors. Larger differences between the thermal conductivities of the carrier gas and the sample will create a more definite detector response. Since larger molecules exhibit lower thermal conductivities, a light carrier gas such as hydrogen or helium should provide the best results. The thermal conductivities of several common gases are listed in Table 11-1.

Other detector parameters should be considered to optimize TCD performance. An increase in filament temperature will increase the output signal. This will also result in a higher filament resistance bringing about a great increase in sensitivity. If the current is boosted too high, noise may develop and there is always the risk of burning out the filament. Detector block temperature is also important. It is heated primarily to minimize temperature fluctuations from the outside air and to prevent condensation of sample vapors. Because sensitivity will increase with a greater difference between block temperature and filament temperature, it is advantageous to keep the block temperature as low as possible. High carrier gas flow rates would result in a lower filament temperature and a corresponding loss in sensitivity. For this reason, lower flow rates are also beneficial.

Because the TCD is a concentration dependent detector it requires good flow control. Fluctuations in the flow rate of the carrier gas may result in a noisy baseline or irregular drift of the baseline. To check the TC cell for leaks, first cap off the exit ports of the detector cells. Then pressurize the detector cells to 30-40 psig and turn off the pressure source. If the pressure remains constant, the cells are leak tight. This same method is useful for leak checking the entire flow system, though small leaks may not be immediately obvious. Extreme changes in the carrier gas flow caused by loose particles trapped in the gas line (a crumbling injection port septum is a typical source) can result in difficulties in zeroing the recorder pen or cause the pen to go completely off scale and not return.

When operating at carrier gas flow rates of 10 cm^3/min or less, it is advisable to install about three to six inches of 0.01 inch i.d. tubing at the detector outlet. This tubing will act as a restrictor to prevent back diffusion of atmospheric air into the detector cell. A similar restrictor is advisable at the detector inlet when backflushing or multicolumn valves are used as part of the GC system. A restrictor in this position will dampen pressure surges into the cell and result in better cell stability.

Filaments in the TC cell require some care. The filament current should be monitored; and, if the current does not remain constant, baseline noise or a gradually drifting baseline will result. If there is no current showing on the meter, the problem is either a meter malfunction or a broken filament. A broken filament will result in the recorder pen

TABLE 11-1

Thermal conductivities and response values for selected compounds

Compound	TC at 100°C (Cal/cm sec deg x 105)	TC relative to He = 100	RMR*
Carrier gases			
Argon	5.2	12.5	--
Carbon dioxide	5.3	12.7	--
Helium	41.6	100.0	--
Hydrogen	53.4	128.0	--
Nitrogen	7.5	18.0	--
Samples			
Ethane	7.3	17.5	51
n-Butane	5.6	13.5	85
n-Nonane	4.5	10.8	177
i-Butane	5.8	14.0	82
Cyclohexane	4.2	10.1	114
Benzene	4.1	9.9	100
Acetone	4.0	9.6	86
Ethanol	5.3	12.7	72
Chloroform	2.5	6.0	108
Methyliodide	1.9	4.6	96
Ethyl Acetate	4.1	9.9	111

*Relative molar response in helium. Standard: benzene = 100. Taken from references 82 and 91. (Courtesy *J. Gas Chromatog.)* (5).

remaining off scale. Filaments can burn out, particularly if highly reactive samples have been run through the detector. To protect the filaments, it is advisable that the filament current knobs be turned down as low as possible before turning the cell on. This reduces the power surge to the filament resulting in a longer filament life. Another precaution which will increase filament life is to be sure that the detector cell has been purged with inert carrier gas before turning the cell current on. This can prevent oxidation of the filaments. Problems which appear to be caused by the filament may only be the result of loose or faulty connections between the detector and power supply (Detector Cleaning - Appendix A).

FLAME IONIZATION DETECTOR

Even though a more recent development than the thermal conductivity detector, the flame ionization detector (FID) is being used more today than any other type of GC detector. The FID is nearly a universal detector responding to all but a few gases, such as the so called "permanent" gases, nitrogen, oxides of nitrogen, H_2S, SO_2, COS, CS_2, CO, CO_2, H_2O, and HCOOH. Actually, this insensitivity can be used to good advantage by using H_2O or CS_2 as a solvent, since large solvent peaks are eliminated. The presence of some of these compounds can have some effect on detector performance however. While other detectors are more universal, the FID has the advantage of being able to detect concentrations as low as 10^{-12} gm/ml.

Linearity of response is also a valuable asset of the FID. While a linearity in the range of 10^4 is not uncommon in detectors, the linearity of an FID has been shown to cover a range of 10^7 (8).

Even though flame ionization detectors are extremely sensitive, they are not particularly complicated. The basic design of the detector is shown in Figure 11-7 (9). An electrical field is established between a negatively charged hydrogen burner jet and a positively charged electrode. Column effluent is ionized in the flame and the negative ions are collected at the detector anode. To insure total ionization of the effluent, an excess of oxygen is introduced into the FID chamber by introducing air at the base of the flame jet.

While a given set of detector conditions can be used for a wide range of GC applications, the detector should be optimized for each column, column flow rate, and sample. A plot of sensitivity vs. hydrogen flow rate as shown in Figure 11-8 (10), will show a maximum sensitivity probably near 30 ml/min. Changes in carrier gas flow rate of not more than 25% should not greatly affect the flame response (11). The flow of air into the detector should be of the order of 10 times the flow rate of H_2. Various other performance factors for FID, such as flame jet temperature and applied voltage are discussed by McWilliam (12).

The FID is subject to some limitations. Since column effluent is destroyed during detection, use of the effluent for further analysis necessitates the use of an effluent splitter. Response of the detector to pure hydrocarbons (containing only hydrogen and carbon) is approximately linear with respect to carbon number; substituted groups, however, will decrease the sensitivity of the detector. For substi-

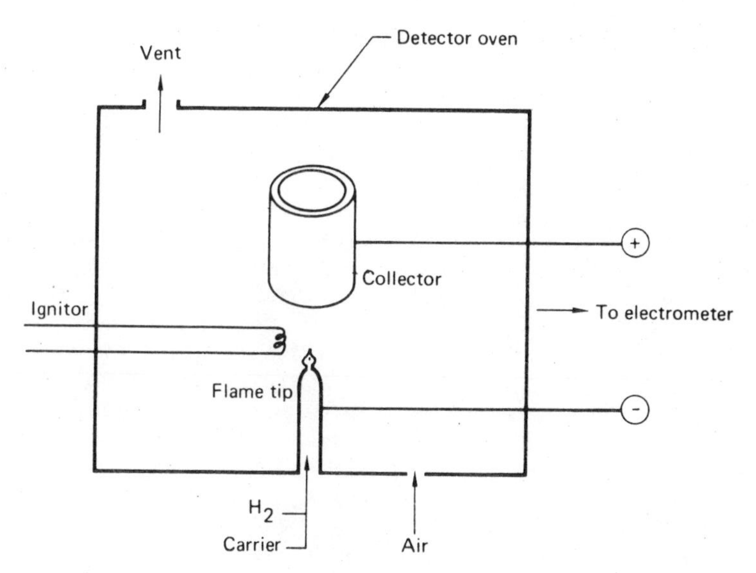

Fig. 11-7. The components of a flame ionization detector (FID) (Courtesy of Anal. Chem.*) (9).*

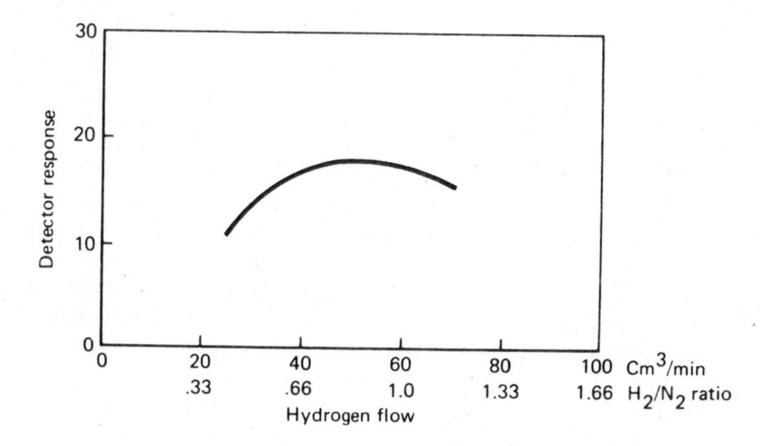

Fig. 11-8. The sensitivity to an FID is affected by the H_2 flow rate of the flame (10) (Courtesy of Hewlett-Packard).

tuted hydrocarbons (substituted with atoms of halogen, sulfur, phosphorous, oxygen, or nitrogen, for example) the FID requires the use of calibration curves for accurate quantitation.

An FID developed by researchers at Shell Oil Company (13) provides approximately 200 times the usable sensitivity of conventional FID apparatus. This is accomplished through changes in the detector geometry and in the applied voltages. An excellent review of factors related to TCD and FID operation has been published by Dietz (14).

Flame detectors are very reliable instruments, however, they too require periodic maintenance. It may be found, for instance, that the ignitor fails to ignite the flame. This can be due to a broken ignitor coil or lack of voltage from the electrometer. Ignitor failure can usually be traced to a defective ignitor voltage cable. Heat at the junction of the detector and ignitor cable can cause an oxidative film to build up, thus preventing a good electrical connection. The junction can easily be cleaned with a piece of emergy cloth or other abrasive. Before igniting the flame, always be sure that the detector temperature is at least 125°C to prevent condensation of water formed from the combustion of hydrogen. Never attempt to light the flame with an ordinary match. Phosphorous and other foreign matter could fall into the detector giving rise to noise. If it is necessary to ignite the flame without the use of the ignitor, construct a small wick by inserting a pipe cleaner into a pyrex glass tube. Dip the pipe cleaner in methanol and ignite it. This is a safe way to light the FID. If the flame ignites but fails to remain lit, check first to see that hydrogen flow and air flow are at their proper levels. If all flow controllers and pressure regulators are in their proper position, check to see if there is a flow coming from the flame tip by connecting the flame tip to a flow meter through a section of teflon tubing (to prevent contamination of the tip). A restriction of the flame tip can sometimes be removed with a piece of thin clean wire. Caution: Before handling the detector in any way be sure to turn off all power sources. A high carrier gas flow rate may cause the flame to blow out. A splitter can be used to decrease the amount of carrier gas reaching the detector if the high flow rates are desirable. Large samples entering the detector can overload it and quench the flame.

It is not difficult to determine whether the source of the noise is from the detector or if the source is elsewhere. First, remove the signal output and cover the input connector with a piece of aluminum foil. If the baseline is still noisy, the problem lies in the electrometer or recorder, not the detector. If a steady baseline results, the source of the noise is probably the detector. Inspect the entire detector for buildup of soot or other extraneous particles and clean it if necessary (see Appendix A). Excessive column bleed which has resulted from poor column or excessive column temperatures can usually be detected by examining the flame. Impurities entering the flame will give the flame color while a normal flame is invisible. Be sure that a pure carrier gas is used and that air entering the detector is filtered as these are both causes of detector noise. If the ignitor coil is improperly positioned above the flame tip, a great deal of noise will result. Since detector geometries vary, consult the instrument manual for proper positioning. A frequent cause of detector contamination is column bleed. If the teflon insulators have become encrusted with deposits, this can be alleviated by scraping them with a sharp knife. Reducing the column operating temperature is the best way to reduce column bleed entering the detector.

An alternate approach is to adapt an adsorbing column between the separation column and the detector. This adsorbing column can be made by filling a short 6-inch by 1/8 inch O.D. section of tubing with a 3% solution of CW20M-TPA (Carbowax 20,000-Terephthalic Acid) liquid phase on a solid support. This column will absorb column bleed for a time, but will not function indefinitely. However, several columns of this type can be prepared simultaneously.

The detector may become contaminated to the extent that it cannot be cleaned simply by raising the detector temperature. It will then be necessary to clean the detector as outlined in Appendix B.

ELECTRON CAPTURE DETECTORS

The electron capture detector (ECD) was developed around 1960 by Lovelock and Lipsky (15). Another type of ionization detector, the ECD uses a radioactive source to generate the ions measured in these detectors. Rapidly moving free electrons from the radioactive source (H^3, Ni^{63}, or Kr^{85}) will be captured by molecules in the detector to form stable negative ions or charged atoms. This reaction occurs by one of two methods:

a) $AB + e^- \dashrightarrow (AB)^- + \text{energy}$
b) $AB + e^- \dashrightarrow A^- + B \pm \text{energy}$

Not all molecules exhibit the same capability to capture the free electrons. This is a measure of the electron affinity of a molecule. Those molecules showing lower capability for capturing free electrons are said to possess a lower electron affinity.

In the EC detector, a carrier gas, usually 95% argon and 5% CH_4, is flowing past an ion source as shown in Figure 11-9. The free electrons ionize the nitrogen creating a current flow. As samples with higher electron affinities enter the detector chamber, the current flow is reduced. Unfortunately, the amount of current reduction is not only a function

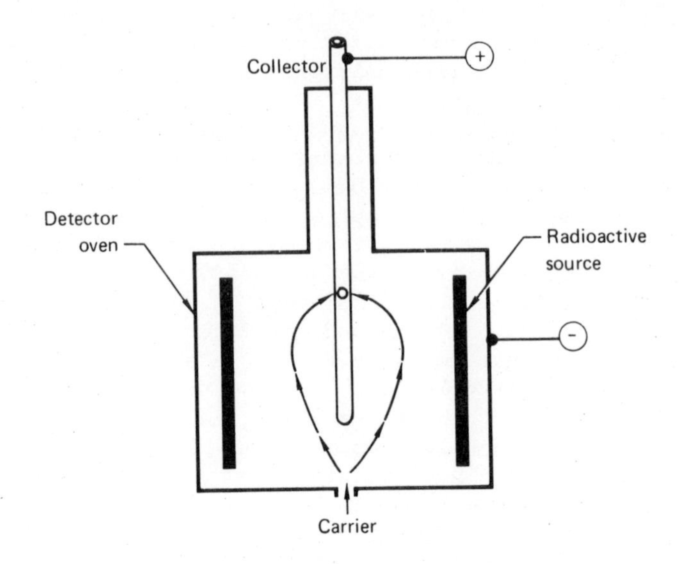

Fig. 11-9. The components of an electron capture detector (ECD) (11). (Courtesy of Anal. Chem.)

of the amount of sample present, but also the electron affinity of the sample. For quantitation then, a calibration must be made separately for each sample component.

The ECD exhibits considerable selectivity. It is highly sensitive to halogenated compounds (I, Br, Cl, F) nitrates, conjugated carbonyls, and certain organo metallic compounds. It is not sensitive, however, to compounds with low electron affinities, such as aliphatics and alcohols. The selectivity of the ECD for halogen compounds has led to its extensive use in the field of pesticide research. For compounds such as lindane and aldrin ECD can detect quantities as low as 0.1 picogram.

The primary source of problems with EC detectors is their low range of linearity. While FID detectors show linearities over a range of 10^7, an ECD cannot exceed 5×10^2. Linearity will vary according to the radioactive source used. Tritium, H^3, provides the widest linear range as seen in Figure 11-10. Ni^{63} provides a linear range of only about 50, but the linearity does not level off sharply as with H^3. Use of a calibration curve can extend the useful range of ECD with an Ni^{63} source to nearly 10^3.

Other factors should be considered in the choice of the radiation source. H^3 is the least temperature stable of the two isotopes and is limited to 220°C by the Atomic Energy Commission. Low vapor pressure samples can condense on the

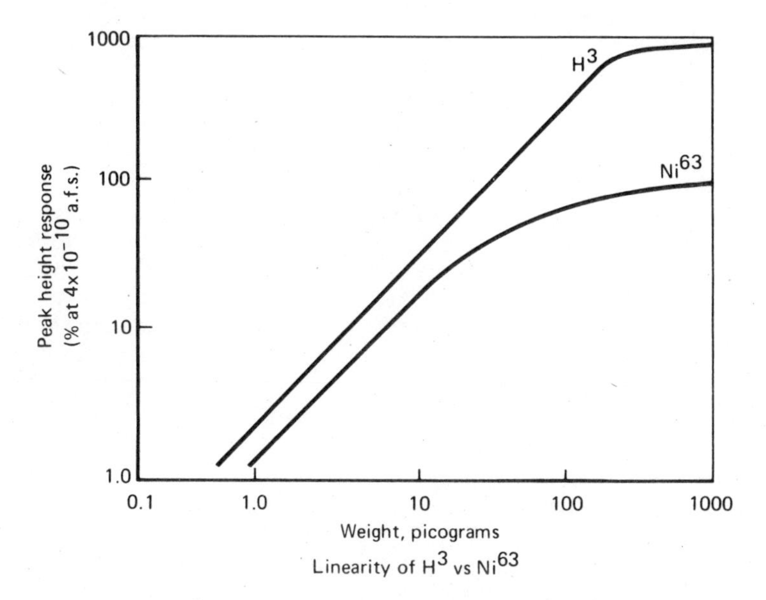

Fig. 11-10. An H^3 source for an EC detector exhibits a wider linear range than does the Ni^{63} source. (Courtesy Varian Aerograph) (17).

tritium (H^3) detector causing harmful contamination. Ni^{63} sources are capable of withstanding temperatures up to 350°C lessening the danger of contamination of the detector and reducing the danger to the operator of harmful radiation. Recently a tritium source has been described with a maximum operating temperature of 300°C (17). This will undoubtedly increase the popularity of tritium sources for ECD detectors.

Many studies have been made concerning optimizing of ECD operation (18, 19). Detector temperature, applied voltage, and carrier gas flow rate will influence sensitivity. A temperature increase will bring about an increase in sensitivity, but an increase in carrier gas flow rate will result in a decrease in sensitivity. Variation in applied voltage also influences sensitivity and can, if desired, contribute to the selectivity of the detector. Since increasing the applied voltage will decrease sensitivity, compounds with low electron affinities can be screened out by a high applied voltage. A pulsed, rather than a continuous voltage is often used to minimize the effects of flow rate and temperature changes.

If the standing current of the EC detector is low, or if the baseline dips negative after peaks, cleaning is necessary. Cleaning requires proper AEC authorization, though the method is simple as outlined in Appendix B.

The ECD requires accurate temperature control. If the temperature is allowed to drift, so will the baseline. AEC regulations require that the unit be equipped with upper temperature limit controls to prevent the detector from rising above the limit for the radioactive source. An exhaust vent above the ECD is also required in the event that the controls do not function properly.

Except for cleaning procedures and detector temperature control requirements, troubleshooting methods for the ECD are similar to those for the FID.

Some problems are involved in the operation of an ECD. Quantitation requires extensive calibration and usually it is difficult to obtain good accuracy over a wide range. Care must be taken to avoid contamination of the detector and a hot septum should be avoided. The ECD is highly sensitive to O_2 (1 ppm) and care must be taken to eliminate any possible air leaks in GC system. Some methods are available for removing small amounts of oxygen and water from the carrier gas stream before it enters the detector. One of these, called the Oxisorb (TM) vessel (manufactured by Analabs, Inc., North Haven, Conn.) can hold oxygen impurities below 0.1 ppm (where inlet does not exceed 15 ppm) and can hold water below 0.5 ppm (where inlet does not exceed 10 ppm). "Wet" samples should be dried in a suitable fashion prior to analysis. Even traces of water in samples cause appreciable noise problems which last for some duration due to adsorption effects within the detector.

ALKALI FLAME IONIZATION DETECTORS

Phosphorus containing compounds are selectively detected by the alkali flame ionization detector. Similar in operation to the FID discussed earlier, the major difference is the addition of a quantity of an alkali salt placed near the tip of the flame, as seen in Figure 11-11 (20). Cesium bromide (CsBr) and rubidium sulfate (Rb_2SO_4) have both been used as the salt in a variety of configurations including pellets, impregnated screen and coated wire loops. A lower air flow rate is also used. Generally, an air flow rate of approximately 130 ml/min is used as opposed to flow rates of 300 ml/min or more for normal FID. It is essential to maintain constant hydrogen and air flow rates for good quantitation. This demands the use of good flow controllers. The stability of hydrogen flow cannot be overstressed. Adjustment in the flow rates can optimize this detector for nitrogen compounds. While no clear theoretical explanation for the operation of the AFID exists at this time, it nonetheless enhances the response to phosphorous compounds by 5000 to 1 and the response of nitrogen compounds by 25 to 1 over

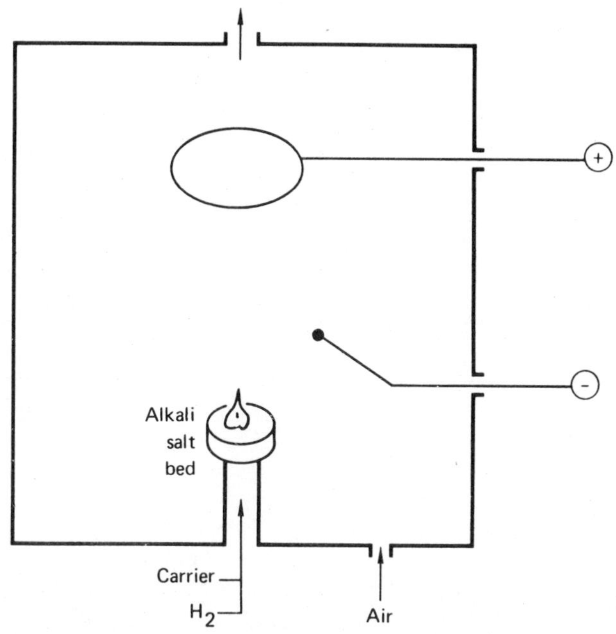

Fig. 11-11. The basic parts of an alkali flame ionization detector (AFID). (Courtesy of Anal. Chem.*) (20).*

the response for pure hydrocarbons.

Some precautions should be exercised in the operation of an AFID. Use of more than 1:1 ratio of sample to solvent will seriously upset the detector response. A considerable amount of time is sometimes necessary for the detector to return to normal. The sensitivity of the detector is subject to gradual changes because of losses from the alkali pellet, which must be replaced periodically. It is therefore, necessary to make frequent calibrations of the detector to maintain accurate quantitative results.

FLAME PHOTOMETRIC DETECTORS

The increasing interest in air pollution has added to the popularity of another type of detector. Flame photometric detectors (FPD) can be adjusted to obtain selectivity for either sulfur or phosphorus compounds, both of which are common constituents of air pollutants. Not only are these detectors highly selective, they are also sensitive enough to detect nanogram quantities of organo-phosphorus compounds.

The basic idea behind the FPD is the measurement of emittance from the light of a hydrogen flame. As shown in Figure 11-12 carrier gas mixed with oxygen rich air enters a

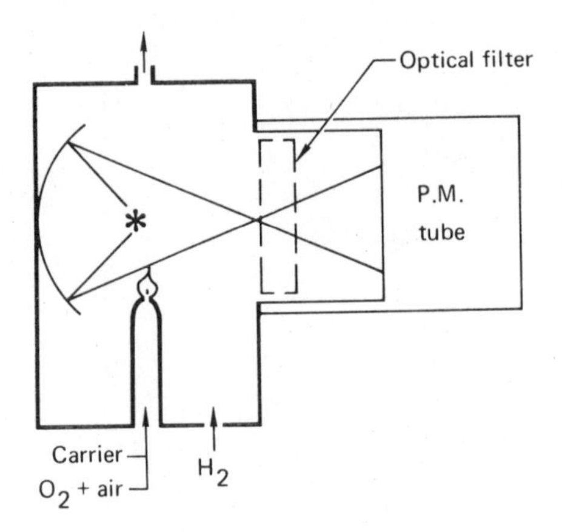

Fig. 11-12. The basic parts of a flame photometric detector (FPD) (20). (Courtesy of Anal. Chem.)

hydrogen filled chamber through a burner tip. Light from the flame impinges upon a mirror and is reflected to an optical filter. The optical filter allows only light of either 526 mμ (for phosphorous detectors) or 394 mμ (for sulfur detectors) to pass through to a photomultiplier tube. Current from the photomultiplier tube is sent to the electrometer.

Aside from its specificity, other factors keep the FPD from more widespread use. First, the FPD exhibits little or no linearity. For accurate quantitation, a calibration curve must be determined for each compound in the sample. Due to the arrangement of the flame and the large amount of hydrogen present, sample solvents are likely to extinguish the flame. In more recent models a manually operated column effluent stream diverter is available to prevent solvent from entering the detector (the new Bendix model). Another universal detector is the cross section detector developed originally by Deisler and co-workers around 1955 (21). While lacking something in sensitivity, these detectors are nonetheless dependable and relatively simple. Beta particles from a tritium source ionize samples entering the detector. A carrier gas is used which has a low cross section (H_2 or He + 3% CH_4) thus diminishing the probability of an ionizing beta particle collision. A potential applied across two closely spaced electrodes will collect any ions formed from beta particle collision. When a sample enters the detector which has a higher cross section than the carrier gas, more collisions will result, producing a higher current. This current increase is amplified and recorded. A typical cross section detector is shown in Figure 11-13 (22).

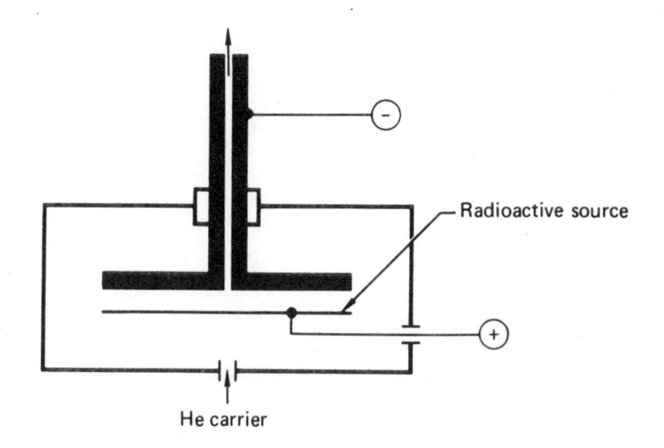

Fig. 11-13. The basic parts of a cross-section detector and a helium ionization detector (HeD); the chief difference between the two being the use of the field gradient with the HeD. (Courtesy of Anal. Chem. *(22).*

The sensitivity of this detector varies somewhat with the cross section of the molecule being detected. Clark (23) showed that the sensitivity could be predicted on the basis of their calculated molecular cross sections. Once again, AEC regulations must be adhered to in the handling of the radioactive source.

HELIUM DETECTOR

Helium detectors (HeD) provide for the detection of minute quantities of permanent gases. These detectors are physically very similar to the cross section detector. The major difference is the utilization of a very high field gradient across the detector electrodes. This causes the He carrier gas to enter a metastable He* state having an ionization potential of 19.8 eV. Only neon, among the permanent gases, exceeds this ionization potential, thus all samples are ionized producing a positive signal.

A HeD must be used only with gas solid chromatography, since column bleed from liquid loaded columns will quench the metastable He* before it can ionize sample molecules. Quantitation in the ppb range is difficult with the HeD and somewhat limited by the lack of an adequate calibration system. Still, HeD is fifty times more sensitive to methane, for example, than the FID. This in itself will harbor to the the continued use of this detector.

THE GAS DENSITY BALANCE (32)

One of the original GC detectors was the gas density balance (GDB). It is not in use much today, but it still has some properties which make it useful. The foremost advantage of a gas density balance is that the column effluent never comes into contact with the detector sensors. This makes the GDB useful for detection of corrosive materials which otherwise might cause damage to other detectors. In addition to this, the GDB is comparatively simple, requires no calibration, can be used with readily available carrier gases (nitrogen, argon, carbon dioxide), sulfur hexafluoride and is non-destructive.

Operation of the GDB can best be described by referring to the diagrams in Figure 11-14 (4). In Figure 11-14a, a reference gas identical to the column carrier gas enters at C. Column carrier gas enters at D. When the gas entering at C is identical to the gas entering at D (i.e., no sample in carrier gas stream), flow past sensors A and B will be identical. Sensors at A and B are heated elements connected through a Wheatstone Bridge.

If, as in Figure 11-14b, the column effluent contains a sample more dense than the carrier gas, downward flow of the sample will retard flow of the reference gas past sensor B. The result is a change in the amount of heat carried away from sensor B causing a change in the element resistance with subsequent imbalance of the Wheatstone Bridge.

It is recommended that high purity carrier gases free of oxygen be used and the normal precautions are used with regard to drying the gases using molecular sieve tubes before entering the system.

The principle of the gas density detector permits the analysis of highly corrosive materials not possible with many other types of detectors. One must also consider the other features in the flow scheme as well, such as the injection port, column and associated transfer lines before introducing corrosive materials.

A more recent development in the use of gas density detectors is that of the Mass Chromatograph, introduced by Chromalytics Corporation and seen in Figure 11-15. This instrument uses dual gas density detectors and perfectly matched GC columns such that the flow characteristics between the two detectors are virtually the same. Two separate carrier gases are used. The more common choices are sulfur hexafluoride, molecular weight 146, and nitrogen, molecular weight 28. This instrument is used primarily for molecular weight determinations of components in a GC sample. Operated in this mode, molecular weight data may be obtained for some samples which, if analyzed by mass spectrometry, may fail to yield a parent molecular ion. Perhaps the weakest link in using this mode is the demand for absolute flow control between the two detectors. The instrument may also be operated in the quantitative mode, just as any other gas chromatograph.

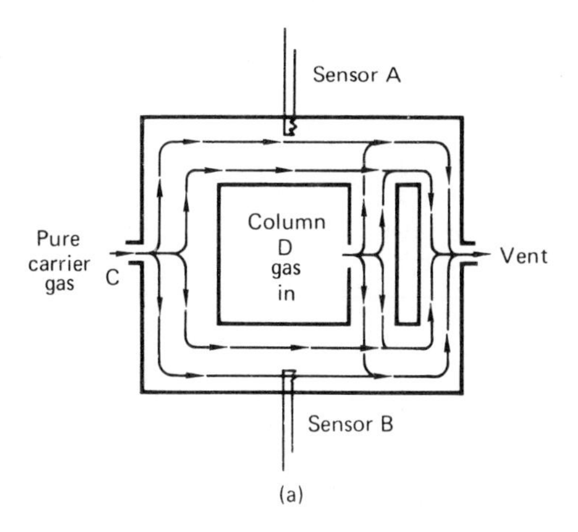

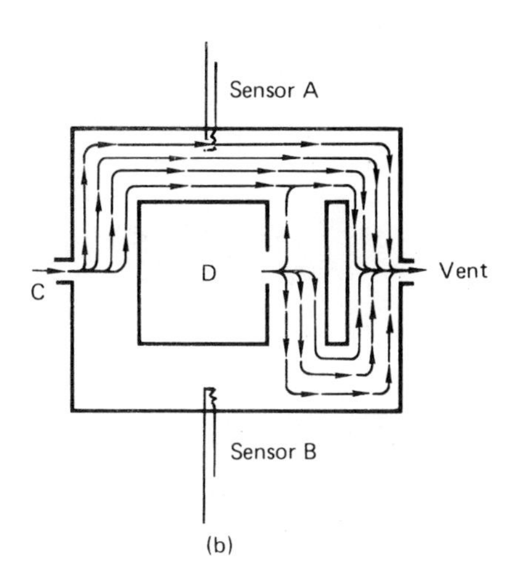

Fig. 11-14. The principles of operation of a gas density balance: (a) no sample from the column, equal flow across sensors A and B, (b) with sample from the column, the flow across sensor B is blocked off forcing more flow through sensor A (25).

COULSON CONDUCTIVITY DETECTORS (32)

Let's briefly turn our attention to the operational and maintenance aspects of the Coulson Conductivity detector.

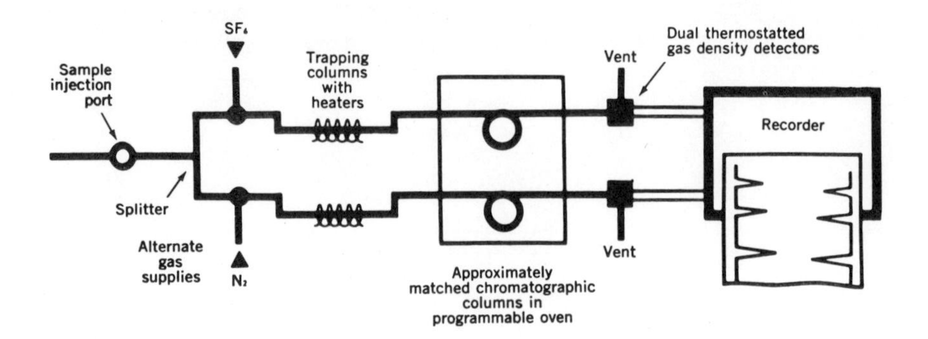

Fig. 11-15. Chemalytics molecular weight chromatograph splits sample between two columns, uses two carrier gases.

Figure 11-16 in the text is a simplified block diagram of this system showing some of the basic applications and principles of operation. This detector is capable of operating either in the *reductive mode* to determine trace organic nitrogen and chlorine compounds, or in the *oxidative mode* to determine trace organic sulfur compounds.

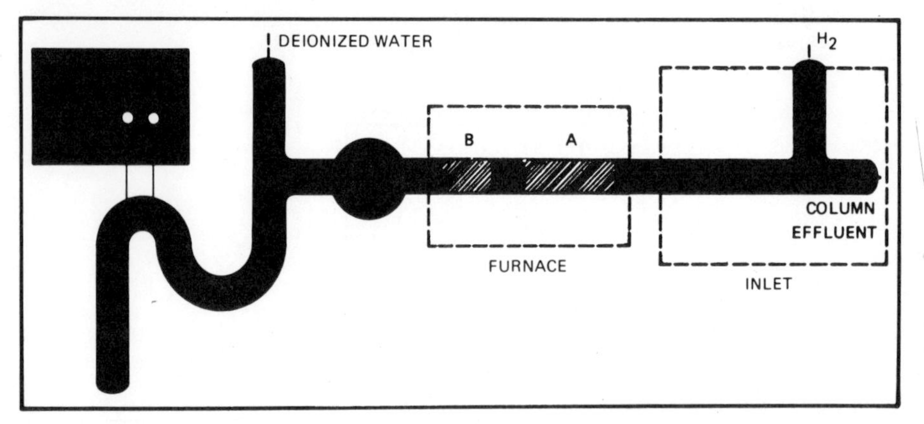

Fig. 11-16. Block diagram of Coulson conductivity detector (33)

In principle, the GC column effluent is mixed with hydrogen or oxygen, depending again upon the mode of operation one chooses, and passes into a quartz tube furnace or pyrolyzer which is maintained at 850 to 900°C. It is in this tube that the organics are converted to detectable species. These converted species are swept out of the pyrolysis tube and are dissolved in a continuous flowing stream of deionized water. The flowing stream sweeps the dissolved products past immersed platinum electrodes also positioned in the flow scheme. Those components which are soluble and have sufficient ionization in water are detected as the electrical conductivity of the flowing stream changes. These changes are measured using a dc bridge circuit and recorder.

Let's refer now to Figure 11-17 in your book which shows a more detailed block diagram of the detector and its associated parts. In the upper left-hand corner you will observe a region designated "vent valve." This valve is extremely useful for venting sample solvents, undesired "heavy ends" in a

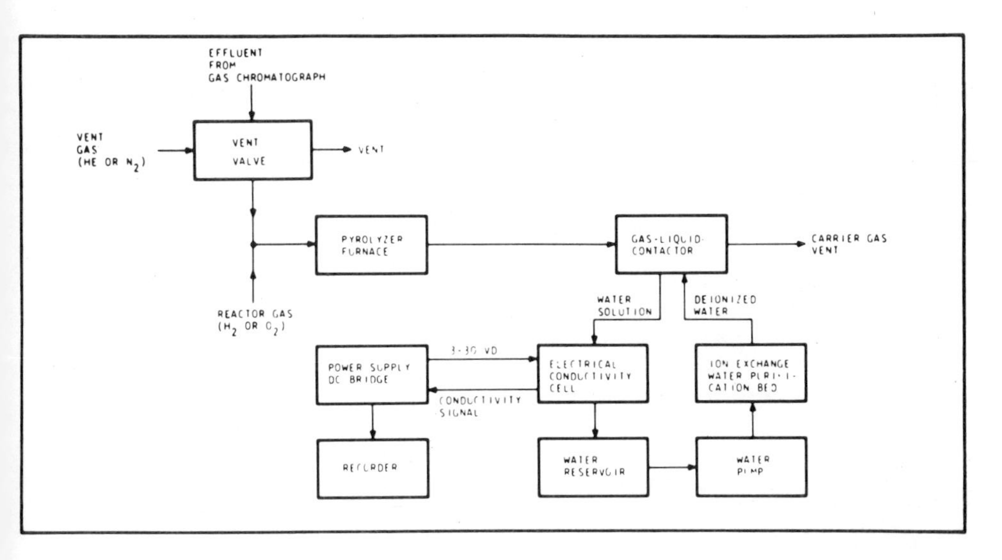

Fig. 11-17. Block diagram, Coulson electrolytic conductivity detector system. (Courtesy of Tracor, Inc.)

sample, etc. A plumbing diagram of this valve is seen in Figure 11-18. Normally, this valve is operated at temperatures of 200-225°C and is manipulated manually using an external lever switch. As with other valves operated at these temperatures, problems do develop so that the valve will stick "open" or "closed" and it becomes virtually impossible to manipulate the lever. Generally, this occurs when the valve has been left in either position for perhaps one or two days without changing its position. Most often, this occurs in a "standby situation" whereby the instrument is idle. In these cases, it is recommended that the vent valve temperature be lowered fifty degrees or so, or alternatively, manipulate the valve periodically to prevent its sticking in one position. For uninterrupted service, either a substitute valve or a replacement should be kept on hand.

Moving on through the diagram shown in Figure 11-17 let's focus our attention now on the pyrolyzer furnace. Certainly, one of the regions of major concern in the entire

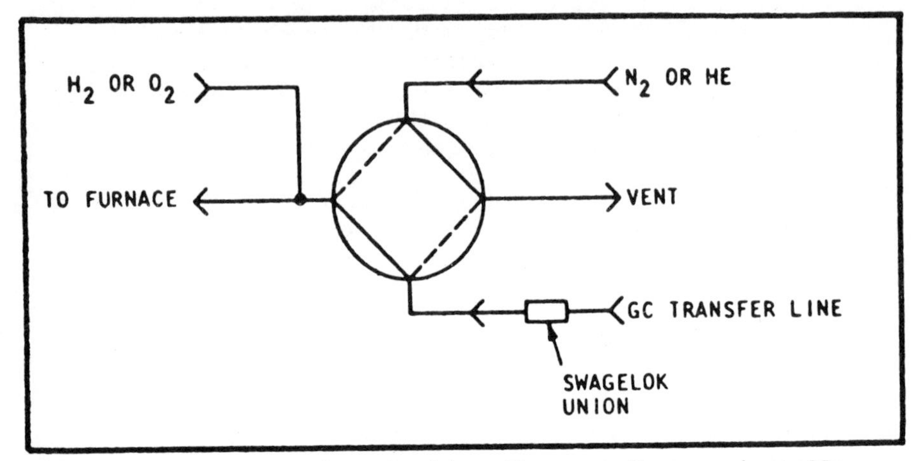

Fig. 11-18. Vent valve plumbing diagram, factory installation. (Courtesy of Tracor, Inc.)

assembly is that of the pyrolyzer tube exit and associated fittings, as seen in Figure 11-19. This assembly provides the interface between the pyrolyzer tube and the conductivity cell. This interfacing is accomplished by way of a modified 1/4 inch to 1/4 inch stainless steel Swagelok union and is securely mounted with a bracket to reduce breakage problems. This interface, although unique in design, is susceptible to some problems. The major problem, of course, is that of leaks in the system. If the teflon tubing transfer line is seated properly and securely through the silicone rubber septum and all fittings are secured, including the ball

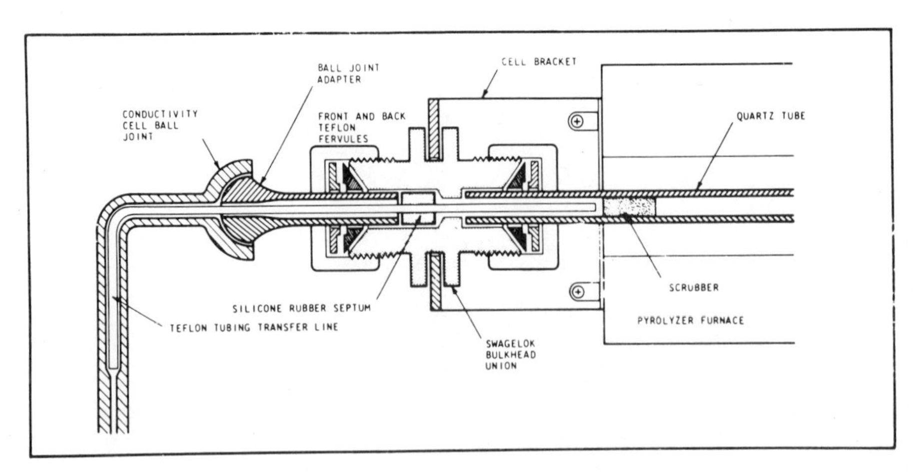

Fig. 11-19. Conductivity cell and bracket assembly. (Courtesy of Tracor, Inc.)

joint adapter, then there should be no problems in the area. We have found that the position and length of the teflon tubing is quite critical for eliminating noise and band broadening. For example, if the inlet end of the teflon tube is extended into the hot zone of the furnace charring occurs, which, being downstream from the scrubber can cause noise problems. Also, the exit end of the teflon tube should be firmly seated into the reservoir. If this is not firmly seated then turbulence occurs in this region which seems to produce band broadening of the chromatographic peaks.

Let us look now at Figure 11-20 and re-orient ourselves with another flow diagram of the system. The particular region of interest at this point is the *vent tube*. This tube, *under no circumstances*, should ever be positioned below the liquid level in the reservoir, otherwise back pressure on the entire system will be produced.

Continuing on, with reference to Figure 11-20, you will notice that, ideally, a continuous flow of water should be passing the electrodes. The cell siphon tube has a small internal diameter. Occasionally, in the region of the gas-liquid separator, a gas bubble will become lodged in the liquid stream flowing past the electrodes. The result of this is that flow through the siphon tube is halted, resulting in no response at all from the detector. The entrapped bubble can be observed by visual examination of the siphon tube. Once the siphon is re-established the response is

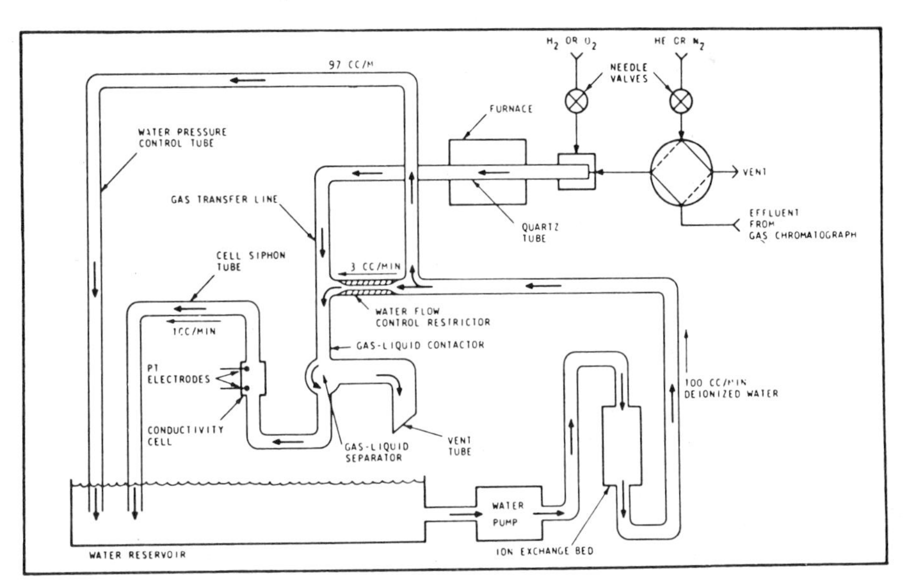

Fig. 11-20. Schematic flow diagram, Coulson electrolytic conductivity detector. (Courtesy of Tracor, Inc.)

restored. The volume of the water level in the reservoir should never be permitted to get below the water pump intake, otherwise the pump will become damaged. For continuous operation, this water level should be checked daily and replenished as necessary.

Referring again to Figure 11-20, the purpose of the ion exchange bed is to maintain and provide a deionized source of water for circulating throughout the system. These resins are quite efficient; however, they must be changed over long periods. A symptom regarding the need to renew these resins is a gradual increase in the noise level as observed on the recorder trace.

Overall care with the glass apparatus is that of having it securely mounted.

Another possible trouble area is the electrical connection between the platinum electrodes and the dc bridge circuit. This is generally accomplished by way of alligator clip leads on the electrodes. If the apparatus is located in an area prone to bench vibration, etc., the clip leads may vibrate loose causing intermittent noise or no response.

In highly contaminated areas, the electrodes themselves may become film-coated resulting in noise or reduced sensitivity. The electrodes should be inspected periodically and, if necessary, buffed with fine emery cloth.

The use of the electrical conductivity detector with older model recorders sometimes presents a problem. Many of these recorders do not possess the necessary noise filtering devices present on newer models. In these instances, that recorder may still be used merely by placing a capacitor across the leads from the bridge circuit to the recorder.

At this point, we have completed the flow diagram of the apparatus. Let's consider now some of the operational modes for which this detector was designed.

Let's refer back to Figure 11-16 again. When operating in the reductive mode for determining trace nitrogen, hydrogen is mixed with the GC column effluent and passes over a nickel catalyst seen as "A" in the figure. A strontium hydroxide or barium carbonate scrubber, designated as "B" in the figure, serves to remove acidic components from the gas stream. Therefore, only nitrogen compounds, which have been converted to ammonia, are detected. In order to obtain reliable data when operating in this mode, the nickel catalyst must be active and the scrubber "B" must be effective in removing acidic components. Obviously, the best assessment of catalyst activity is to periodically analyze a standard solution of known response. It is worthy of note that mere injection of ammonia will not provide this measure as no catalytic conversion is necessary. A symptom of catalyst deactivation is a rather gradual decrease in the response of a known standard solution. Finally, the response becomes inadequate and it is necessary to re-activate the catalyst. To do this, we must disconnect the pyrolysis exit from the conductivity cell ball joint, as shown in Figure 11-19. After this is done, the hydrogen inlet gas is disconnected from the vent valve and replaced with compressed oxygen, as shown in Figure 11-18. The pyrolysis tube furnace is maintained at its normal operating temperature and the oxygen supply is turned "on." Oxygen is permitted to flow over the catalyst for ten to fifteen minutes and discontinued. During this period, the scrubber may also be replaced with fresh material. The hydrogen is then re-connected and the pyrolysis exit tube is repositioned for normal operation. Generally, 15 minutes is adequate for the system to re-equilibrate.

The determination of organic halides in the reductive mode presents no additional problems. No catalyst nor scrubbers are used in the pyrolysis tube for this mode of operation.

In the oxidative mode for the determination of organic sulfur compounds, oxygen is introduced at the vent valve location rather than hydrogen, as shown in Figure 11-18. Additionally, selectivity is attained by insertion of a silver wire scrubber which removes the halides formed in this mode. No catalyst is used.

Liquid stationary phases should be carefully selected when using this detector. Use no liquid phase containing the components being sought. For example, do not use polyethyleneimine as a liquid stationary phase when operating in the reductive mode for trace organic nitrogen compounds. Unusually high background with subsequent sensitivity losses are obtained in this case.

SUMMARY

Because of its versatility and high sensitivity, FID is the most widely used detector today. Close behind is the TCD which is one of the early detectors used in GC. In the future it is probable that more work will be done on the use of spectrographic methods of detection. Currently, the major problems in this area lie in interfacing techniques and the high cost of infrared and mass spectrographic equipment. Optimum utilization of current detector technology, however, can provide a great deal of accurate and meaningful information. Table 11-2 summarizes the characteristics of several GC detectors. Detector cleaning is described in Appendix A.

TABLE 11-2

Comparison of Parameters of GC detectors

Parameter \ Detector	Thermal Conductivity	Flame Ionization	Alkali Flame Ionization	Electron Capture	Cross Section	Gas Density	Flame Photometric	Helium Ionization
Linear Dynamic Range	10^4 to 10^5	10^6	10^3	10^3	3×10^5	10^5	500	5×10^3
Minimum Detectable Quantity, gm/sec	10^{-9}	3×10^{-12}	10^{-11}	3×10^{-14}	3×10^{-11}	10^{-10}	2×10^{-12}	4×10^{-14}
Minimum Detectable Conc.,Parts by Vol	10^{-5}	10^{-11}	10^{-9}	10^{-13}	2×10^{-6}	10^{-6}	4×10^{-10}	10^{-13}
Carrier Gas	H_2 & He	He, N_2	He, N_2	Ar, CH_4 & N_2	H_2 & 5% CH_4	SF_6	N_2	Pure He
Detector Vol, ml	10^{-2} to 1.0	Ca. 10^{-4}	Ca 10^{-3}	0.8	8×10^{-3}	0.3	ca. 10^{-3}	0.3
Applicability to Column Sizes	Packed & Capillary	Packed & Capillary	Packed & Capillary	Packed	Packed & Capillary	Packed	Packed & Capillary	Packed
Applicability to Gases & Liquids	All	Organic Compounds	Phosphorous & Halogen Cpds	Oxygen & Halogens	All	All	Sulfur & Phosphorous	All
Utility in Trace Analysis	Low to Medium	Extremely High	Extremely High	Extremely High	Medium	Medium	Extremely High	Extremely High
References	5	12	16	19,20	21	22	23,24	22

REFERENCES

1. J. MacDonald, McDonnell Douglas Corp., Private Communication, 1971.

2. J. Q. Walker, *Petroleum Refiner, 44,* 34 (1965).

3. Keithley Instrument, Inc. *Technical Bulletin,* p. 2, 1967.

4. J. MacDonald, McDonnell Douglas Corp., Private Communication, 1971.

5. A. E. Lawson, Jr., and J. M. Miller, *J. of Gas Chromatography, 4,* 273 (1966).

6. *Ibid., 4,* 274, 1966.

7. *Ibid., 4,* 276 (1966).

8. L. Ongkihong, in *Gas Chromatography 1960,* R.P. W. Scott, Ed., Butterworths, London (1960), p. 46.

9. C. H. Hartman, *Anal. Chem., 43,* 113 (1971).

10. F & M Scientific Corp., Pre-print No. 1, Pittsburg Conference of the ACS, 1966.

11. C. H. Hartman, *Anal. Chem., 43,* 114 (1971).

12. I. G. McWilliam, *J. Chromatography, 51,* 391 (1970).

13. B. O. Prescott, H. L. Wise, and D. A. Chestnut, U. S. Patent No. 3, 451, 780 (1969).

14. W. H. Dietz, *J. Gas Chromatography, 5,* 68 (1967).

15. J. E. Lovelock and S. R. Lipsky, *J. American Chemistry Society, 82,* 431 (1960).

16. C. H. Hartman, *Anal. Chem., 43,* 117 (1971).

17. C. H. Hartman, D. Oaks, T. Burroughs, *Technical Bulletin* 130-66, Varian Aerograph.

18. D. C. Fenimore and C. M. Davis, in *Advance in Chromatography,* Ed., A. Zlatkis, U. of Houston, Houston, Texas, 1970, p. 130.

19. M. Scolnick, *Technical Bulletin 132-67,* Varian-Aerograph, 1967.

20. C. H. Hartman, *Anal. Chem., 43*, 118 (1971).

21. P. F. Deisler, K. W. McHenry, Jr., and R. H. Wilhelm, *Anal. Chem., 27*, 1366 (1955).

22. C. H. Hartman, *Anal. Chem., 43*, 120 (1971).

23. S. J. Clark, *Gas Pipe*, No. 1, Jan. 1963, Jarrell-Ash Co., Waltham, Mass.

24. A. J. Martin, Technical Bulletin, *Mass Chromatograph*, Chemalytics Corp., 1971.

25. J. MacDonald, McDonnell-Douglas Corp., Private Communication, 1972.

26. D. M. Coulson, *J. Gas Chromatography, 3*, 134 (1965) , 285 (1966).

27. J. F. O'Donnell, *American Laboratory*, Feb. 1969.

28. TRACOR Manual, Tracor Inc., Austin, Texas, 1971.

29. C. L. Guillemin and F. Auricourt, *J. Gas Chromatog., 2*, 156 (1964).

30. A. J. Martin, Technical Bulletin, *Mass Chromatograph,* Chemalytics Corp. 1971.

31. J. Q. Walker, McDonnell Douglas Corp., Private communication, 1972.

32. M. T. Jackson, Private Communication, 1973.

33. TRACOR Instruments, 1973.

QUESTIONS AND ANSWERS

Question: What effect does carrier gas choice and flow rate have on flame ionization sensitivity?

Answer: The effect is large. Figure 11-21 is a summary of operating conditions for optimum response of a flame detector. Figure 11-21 shows the response curves with Ar and N_2 carrier gases are much higher but sharply peaked, as opposed to the lower relatively flat curves with CO_2 and He. The analyst must use precise carrier gas flow controllers for satisfactory repeatibility. The response factors calculated for one detector will not be the same for others.

Question: At exactly what point (distance) in the flame ionization hydrogen supply line should a "snubber" or restrictor be installed for higher carrier gas flow rates?

Answer: A ten-inch by 0.01 inch i.d. restrictor should be as close to the flame jet as practical. Usually about 2-3 inches away from the jet.

Question: How do you handle the SiO_2 problem in the FID when using silane derivatives?

Answer: Inject about 5 ml of "Freon" 113 several times a day to prevent SiO_2 build-up.

Question: How do you reduce noise in the flame photometric detector system?

Answer:
1) Use a post column (1/8 inch O.D. x 6 inch long) as a clean-up column.
2) Use higher purity gases.
3) Check for vibration and thermal noise.
4) Early photometric tubes were not light-tight and subject to noise from light leaks. Cover the tubes with a cylinder painted black on the inside and outside.
5) Dirty filters. Noise is greatest because it is a secondary reaction. So replace or clean with absolute methanol.

Chapter 12

INTEGRAL ELECTRONICS

One of the aspects of GC instrumentation which is best understood in both a qualitative and quantitative sense is the electronics section of the instrument. With only the most fundamental knowledge of electronics, the gas chromatographer can understand the operation of at least two of the basic components present in many gas chromatographs, i.e., the electrometer and recorder. With the advent of capillary columns, more sensitive detectors were required because of the small sample sizes which must be injected into the columns. In addition the use of gas chromatography for analysis of trace components requires an ionization detector (1). The use of flame ionization (FID), argon ionization, electron capture, and cross-section detectors requires an electrometer amplifier for presentation of the detector output on a strip chart recorder. The quantitative analysis of multi-component blends could require much longer analysis times if it were not for automatic digital integrators. While the use of the ionization detectors and automatic digital integrators is by no means limited to analyses with capillary columns, it is with this type of analysis that this combination is most useful.

This chapter is intended to cover the basic operating theory and maintenance techniques of T.C. power supplies, electrometers, recorders, integrators, and to describe various kinds of data-handling techniques. Since the material is intended for the practicing gas chromatographer, only fundamental theory will be covered. Operating procedures for electrometers, recorders, and integrators will be covered, along with potential sources of trouble and possible corrections. With only a fundamental knowledge of the operation of these systems, diagnosis of many of the electronic problems can be simplified.

ELECTROMETERS

The first link in most GC integral electronic systems is the electrometer. The signal output from ionization detectors, whether it be flame, argon, electron capture, or cross section is an extremely small electrical current, i.e., a flow of electrons. If this small current flows through a high resistance element, a significant voltage is developed across the resistor. The current, resistance, and voltage are related by Ohm's law:

$$E = I \times R,$$

where E is the voltage, I is the current, and R is the resistance in units of volts, amperes, and Ohms respectively. If a minute electron current of 10^{-11} amps flows through a resistance of 10^{10} Ohms, the voltage across the resistor must be 0.1 volt or 100 millivolts, as given by Ohm's law. This is more than enough voltage to drive a 1 mv. potentiometric recorder full-scale. However, potentiometric recorders can only accept signals across a resistance of several thousand ohms. It is the function of the electrometer-amplifier to develop an identical voltage across the relatively low resistance elements of the attenuator. A potentiometric recorder (to be discussed later) can then be connected across the attenuator for graphic presentation of the detector output signal.

Figure 12-1 is a block diagram of a typical ionization detector tube-type electrometer hook-up. For purposes of illustration the FID has been chosen; however, the same principles would apply to other ionization detectors. An operational amplifier is connected across the output of the electrometer tube. As the electrons flow into the grid, this pushes the grid to a more negative potential. In doing so the plate potential becomes more positive (a fundamental characteristic of triodes) and a greater demand for electrons is created. The plate current must come from the grid and hence the grid potential would be swung more positive if electrons were not supplied to it from the amplifier. In order to supply these electrons, the amplifier must develop a potential (voltage) across the range (or input) resistor to "pump" electrons to the grid. The amplifier voltage

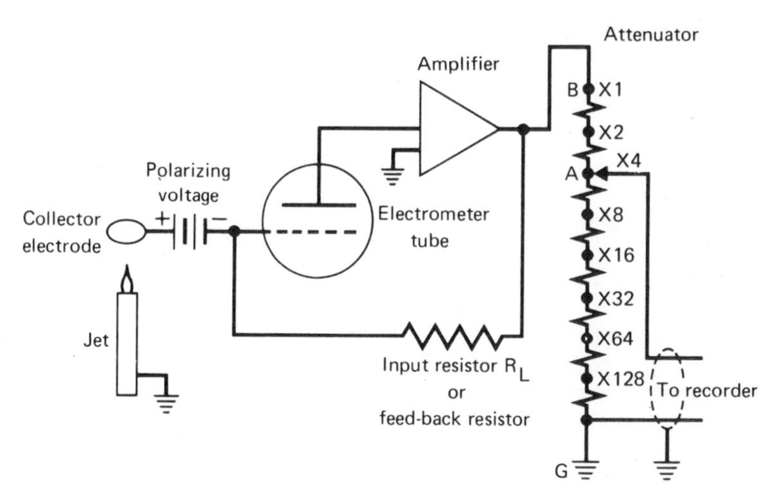

Fig. 12-1 Basic electrometer configuration (Courtesy Varian Aerograph) (3).

output signal (V_a), input resistor (R_L), and "feedback current" are related by Ohm's Law:

$$V_a = i \times R_L.$$

The output voltage (V_a) is impressed across a several thousand ohm resistance attenuator. The voltage drop across the resistance elements of the attenuator circuit can be "picked off" at points corresponding to the various stages of attenuation.

The amplifier, as mentioned, must develop a voltage, hence a negative feedback current to the input circuit is needed. The use of the negative feedback current is necessary for three reasons:

a) To reduce the time constant of the electrometer. The time constant of the input circuit is the lag in response between the input and the output signals. It is determined by the product of the input resistance, which may be 10^{11} ohms, and the capacitance of the input cable from the detector to the electrometer. The capacitance may be 5-25 µµf per inches of cable. The product R x C, in units of seconds could be significant and could affect the response of the system. The effect of the negative feedback is to reduce the time constant by the gain of the amplifier. The gain may be as high as about 10^4.

$$\text{Time constant (secs.)} = \text{Input resistance (ohms)} \times \frac{\text{Capacitance (farads)}}{\text{Amplifier gain}}$$

In this way the time constant is minimized.

b) To operate within the linear region of the electrometer tube. If the ionization current from the flame were allowed to drive the grid potential up and down without limit, operation of the tube could move into a region of non-linear response. This non-linearity would also be reflected in the amplifier output to the recorder. In addition the saturation point of the tube could be reached; beyond this point any increase in the input signal would not be reflected in the output signal. The use of negative feedback maintains operation of the electrometer tube in the linear region of the curve.

c) To maintain a constant detector jet bias potential. The polarizing voltage of the collector electrode is the voltage of the battery *minus* the voltage drop across the input resistor, a variable quantity. Each time a component passes through the flame the jet bias would change and non-linear operation would result if feedback were not used

The input resistor and the attenuator network are the means by which we vary the sensitivity range of the electrometer. The input resistors are in decade steps and the attenuation is normally in binary steps. The attenuator can also be separate from the electrometer, in series with the recorder. The higher the resistance of the input resistor (load resistor, feedback resistor), the greater will be the voltage which the amplifier must develop in order to drive a given current to the electrometer tube grid. The glass envelope of the resistor does not have an infinite resistance and thus sets a limit to which the resistor can be pushed. If the envelope gives less resistance to flow than the resistor wire, the electrons would flow through the path. The grease from a fingerprint on an input resistor can provide a path of less resistance into electron flow. For this reason all parts of the input circuit, especially the input resistors, should *not* be handled by the glass envelope. If they must be handled, use the external leads and preferably, with a pair of pliers. Accidental paths for leakage can lead to a drastic loss in sensitivity. The electrometers used in gas chromatographs have various means to indicate the level of sensitivity or the input resistor selected. For example, Hewlett Packard uses ranges 1, 10, 10^2, 10^3, and 10^4, with range 1 being the most sensitive (2). Varian-Aerograph uses ranges 0.1, 1, 10, and 100 (3). The most sensitive range is the one which includes the highest input resistance, as is needed to produce measurable IR drops from the very small currents produced by ionization detectors. The attenuator is the second means by which we change sensitivity (sometimes an integral part of the electrometer, sometimes separate). The attenuator range may cover a span of from x1 (most sensitive) to x 1024, in binary steps. Some instruments have a ternary attenuator (x1, x3, x10, x30, etc.). The attenuator is the means by which we ordinarily reduce sensitivity to the recorder during the course of a run. Generally, the attenuator is used only to reduce output to the recorder and keep it on scale. Most electrometers can handle the input from GC analysis on one current range without saturation due to a wide dynamic range.

Figure 12-2 is a block diagram for a new solid state electrometer available from Hewlett-Packard, which operates in essentially the same manner as the unit in Figure 12-1, only using solid-state components instead of tubes (2).

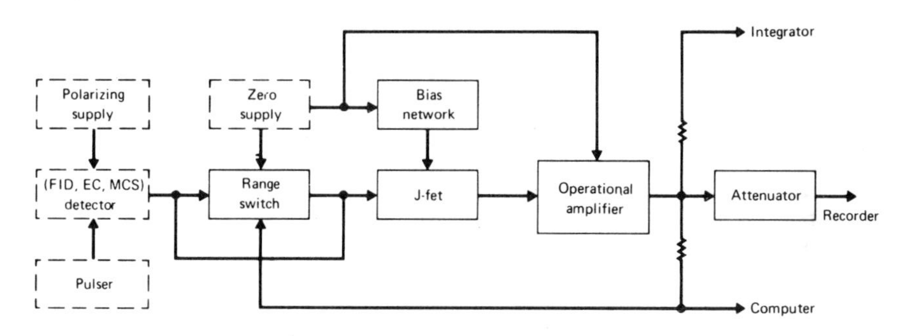

Fig. 12-2. Solid-state electrometer functional schematic (Courtesy of Hewlett-Packard) (2).

FUNCTIONAL OPERATION OF THE ELECTROMETER

There will normally be eight different control knobs on the electrometer. They are: a) mode control, b) detector input, c) balance, d) polarity, e) range, f) coarse, g) fine, and h) attenuation (if attached). Many electrometers will also have a control for a D.C. pulser built into the system. In the use of the electron capture detectors most workers have found that a pulsed D.C. potential on the collector electrode is better than a constant D.C. potential; hence, this section will be used only if the electron capture detector is used. In beginning the operation of the electrometer the first step is to turn the mode control to the warm-up position. In this position voltage is applied to the filament, but no potential is applied to the grid or plate of the electrometer tube. A warm-up period of 15-60 minutes is usually required. Under normal operating conditions the instrument should be left in the warm-up position, but should be turned off for long periods of non-use. After the recommended warm-up period, the mode control is turned to the appropriate position corresponding to the ionization detector being used.

The first point to be considered under "functional operation" is adjustment of recorder zero. Although this is independent of electrometer operation, this step is performed first, with the electrometer disconnected from the recorder (or at infinite attenuation). Recorder zero means that the recorder pen is at zero while the recorder is operational, but no input signal is applied to it. This adjustment is necessary so that signal attenuations will be linear. To insure that no signal is getting to the recorder the attenuator is indexed to infinity or the leads to the recorder are removed. The recorder is zeroed by the zero control on the recorder amplifier. For example, the Honeywell Model 15 has a black dial in the center of the amplifier section for this adjustment. The Honeywell Model 16 control is on the right side along with the span, gain, and damping controls. Rotation of this control moves the pen into coincidence with the zero of the chart paper and scale. The recorder zero drift is usually not large and this adjustment may need to be performed only weekly (or less often). This is also a good time to determine whether the scale zero and chart paper zero are in alignment. The next step is to balance the electrometer. The electrometer output is fed to the recorder, with the detector disconnected from the electrometer. With the attenuator at x1, the balance control is turned until the recorder pen is again at zero. In essence we are balancing the two matched electrometer tubes so that there is no output to the recorder in the absence of a signal input to the electrometer. If the instrument being used is equipped with two or more detectors (dual flame, flame-electron capture, etc.) the detector input selector switch is now turned to feed the desired signal into the electrometer. The instrument is now switched to the desired range. For a flame ionization detector with a small bore capillary column and a sample size of, for example, 0.5-1.0 µl split 100/1, a range of 10^9 amperes is a good one to start with. For a packed column and a sample size up to 10 µl with no split, a range 100 to 1000 times less sensitive is needed. The coarse and fine zero controls are now used to bring the pen to zero again, bucking out any background signal (these controls are sometimes labeled "bucking voltage"). The polarity switch is used to obtain the proper polarity on the bucking voltage. The buck-out operation is ordinarily performed at an attenuation more sensitive than will be used for the analysis.

PROBLEMS AND SOLUTIONS

Following are some of the symptoms that may be observed when the electrometer malfunctions. Since the detector-electrometer-recorder combination produces the chromatogram, the observed symptom may be in any one of the three components and isolation of the trouble is necessary. By electrically isolating each component the trouble can be located. A weak signal response may simply be due to selection of the wrong input resistor setting or attenuation. It may also be due to a shorted input resistor caused by dirt or grease on the resistor envelope as mentioned previously. If it is impossible to balance the electrometer in the absence of a signal from the detector (input cable disconnected or range selector switch in the balance position) the trouble probably lies in the electrometer tubes and they should be replaced.

Loss of signal may also be due to weak batteries in the collector electrode biasing circuit. This voltage is usually kept from 300-350 volts, and weak batteries will not produce efficient collection of electrons.

If the recorder baseline is stable without electrometer output, reconnect the electrometer to the recorder (or change to an attenuation other than infinity) and observe the recorder. If noise is observed at this time the problem must be in the electrometer (if no input current is coming from the detector). If no noise is observed, disconnect the electrometer input cable at the detector, cover the end with foil to eliminate pick-up of stray signals, and observe the recorder. Noise can be produced if the input cable is not properly shielded and grounded. The magnetic field associated with alternating current can act like a generator and cause electron flow in the input circuit. Most input cables are provided with a briaded shield which is attached to a grounding point. Any current which is generated by an external alternating current or changing magnetic field will flow in the shield to ground and will not show up in the output. If no noise has been observed to this point, the problem lies in the detector and/or associated parts.

Drift can be caused by improper stabilization of the electrometer, improper function of the flow control system of the chromatograph, or elution of high boiling components from the column. Although the FID is generally insensitive to changes in flow rate, drift will occur if the change in flow rate brings more or less ionizable material into the FID. Disconnect the detector input to the electrometer. If drift continues, the problem lies in the electrometer/ recorder system and the isolation of the faulty component may be done as described earlier in this chapter.

Occasionally it will be observed that attenuation of the signal will not reduce the signal level by exactly one-half (for a binary attenuator). This is not the fault of the electrometer, but is caused by improper adjustment of recorder zero. If the recorder zero is below the scale zero, attenuation will apparently cut the signal by more than one-half, and conversely if the recorder zero is above scale zero, by less than one-half.

There are two other points which should be mentioned with regard to electrometer operation. The amplifier has a certain limit regarding the output voltage which it can deliver. This will probably be in the range of 10 volts for most amplifiers. If the input signal requires the amplifier develop more than this limit to drive the feedback current to the grid through the input resistor, the amplifier will saturate. Any signal requiring more than this limit will not be reflected in a greater deflection of the recorder pen. This is indicated by flat-topped peaks at some point less than full scale deflection (usually on some high attenuation setting). This should not be confused with the flat-topped peaks which result when the recorder is driven full scale and is mechanically stopped. If saturation is suspected, switch to a higher attenuation to keep the peak(s) on scale and check for flat-topped peaks. This situation will ordinarily be encountered only on the most sensitive ranges; the higher resistance value requires a higher voltage for a given current. Since the amplifier has a limit of 10 volts, any signal greater than 2×10^{-10} amperes would result in saturation and a square peak at about 75% of full scale. In such a case a lower range position should be used. The second point is that the full potential developed by the amplifier is not dropped across the variable portion of the attenuator. In addition to a fixed resistance of 3,750 ohms in the attenuator, an additional resistance element is used in the attenuator network so that only a small per cent of the amplifier output is dropped across the variable portion of the attenuator. The use of the large fixed resistor in the attenuator output does not affect the sensitivity of the system. The attenuator setting is thus the fraction of this small percentage which is being used to drive the recorder.

FUNCTIONAL OPERATION OF THE THERMAL CONDUCTIVITY DETECTOR

An understanding of Ohm's law is also used to see the origin of the voltage developed by a thermal conductivity detector (TCD). Although an electrometer is not used with the TC detector, a knowledge of the electronics involved should be helpful in diagnosing problems with TCD cells.

A typical thermal conductivity detector-Wheatstone Bridge circuit is shown in Figure 12-3. The recorder is connected so that the potential difference between B and C is being measured. The variable resistance in the "current adjust" section is used to vary the total current (I) flowing through the system. This, in part, determines the sensitivity of the set-up. The "zero adjust" control divides the total current (I) through the upper and lower branches (I_1 and I_2). Due to the conservation of electrical charge (Kirchoff's First Law), $I = I_1 + I_2$, the zero control is used

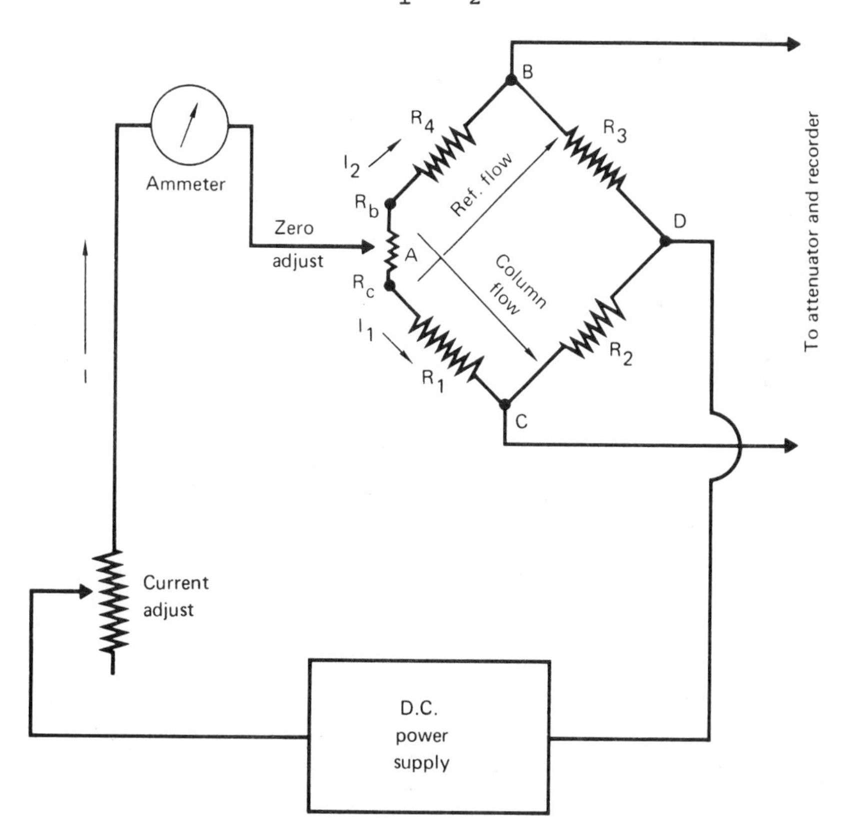

Fig. 12-3. Typical thermal conductivity detector - Wheatstone Bridge circuit (9).

to vary I_1 and I_2 so that the potential drop AB equals the potential drop AC at the null point, and points B and C are at the same potential (no voltage difference-bridge output = 0). Resistances R_3 and R_4 (reference side of TC cell) are constant: R_1 (column side of TC cell) and R_2 are constant also, but can be variable when a sample passes by R_1. The sum of the resistance in each branch ($R_b + R_3 + R_4$ or $R_c + R_1 + R_2$) is thus variable, but equal. Hence, at the null setting I_1 and I_2 are constant. As a component is eluted, R_1 usually increases. Since the potential at B is constant (I_2R_4 constant) and the potential at C is $A-I_1R_1$ (due to change in R_1), a voltage difference is developed between B and C. A recorder connected between B and C can then be used to measure the potential difference. The two electrical factors which affect the sensitivity are I_2 (hence I) and the temperature coefficient of resistance of the sensing elements. These are not mutually independent; a higher current would produce a greater response, but a hotter filament would result which would produce a smaller temperature change for a given amount of heat and filament life would be reduced. For problems and solutions, see Chapter 11.

RECORDERS

The second link in our data system is the potentiometric recorder. It is the function of the recorder to reproduce in a graphic form the output of the electrometer-amplifier or thermal conductivity detector. Figure 12-4 is a block diagram of the components of a continuous balance potentiometric recorder. The 60-cycle line voltage and the chart drive mechanism are independent of the measuring circuit. Since a chromatogram is a presentation of response versus time, the only purpose of the chart drive system is to provide a continuous movement of chart paper in some known relationship of inches of paper per unit time.

The differential amplifier compares the input signal from the amplifier-attenuator network with a reference voltage generated in the measuring circuit of the recorder. The difference between the two signals is called the error signal. This error signal is converted from D.C. to A.C. and amplified to a voltage large enough to drive the servo balancing motor. If the difference is positive the motor is driven in one direction and if negative, in the opposite direction. The balance motor is mechanically linked to the reference slidewire and to the recorder pen. The change in the slidewire contact position changes the value of the reference voltage in such a way as to reduce and maintain the error signal at the amplifier at zero. In essence the system

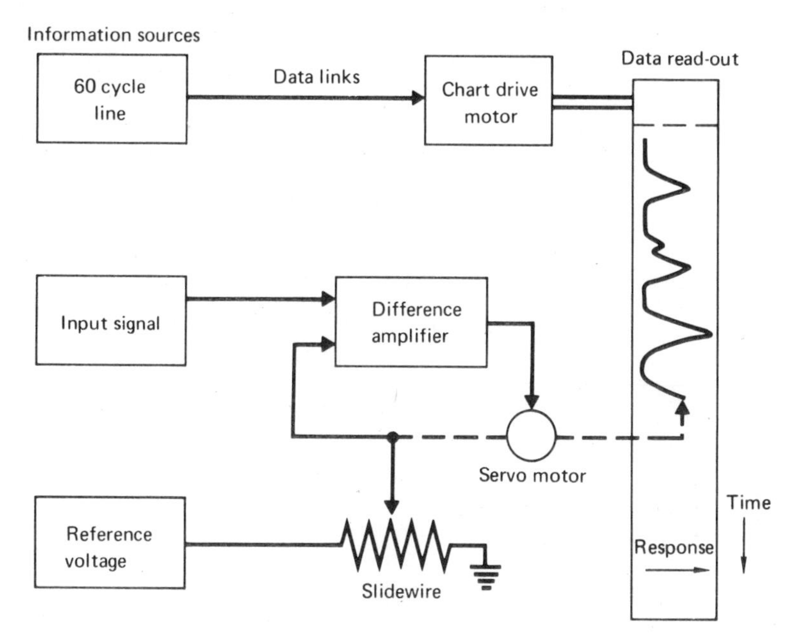

Fig. 12-4. Block diagram of automatic potentiometric recorder (9).

continuously attempts to maintain the error signal at zero. The D.C. voltage across the slidewire is the range of the recorder, for example 1, 5, 10, 50 mV. If the input signal is greater than the reference voltage span, the recorder will be driven full scale and will be mechanically stopped.

The accuracy of the measurements of potential in the millivolt range depends on a precise current flow of known magnitude through the calibrated slidewire (Ohm's law). This constant current supply is provided by a constant voltage unit in the recorder (usually a mercury battery). The constant voltage supply sometimes consists of an A.C. to D.C. power supply and a very accurate voltage regulator. Even in the presence of fluctuations in the line voltage the regulated voltage supply will provide the proper current output to maintain the correct milli-voltage drop across the slidewire. Another component of the recorder is the measuring circuit. This section permits the selection of the span of the recorder, electrical zero, and damping adjustments. Discussion of these points will be given in subsequent paragraphs. The amplifier section of the recorder consists of the components necessary to convert the D.C. input into A.C., amplify the A.C. signal, and feed the output to the balancing motor. The chart drive system consists of a constant speed motor and gear network to move the chart paper at a constant rate. Thus, in Figure 12-4, the measuring circuit unit and the constant voltage source would be part of the reference voltage-slidewire combination and the other essential parts are as given in the figure.

Consideration now is given to some of the terms which are used to characterize recorders. An understanding of these is important in the maintenance and troubleshooting of recorders. The terms "range" and span of a recorder are related. The algebraic difference between the end-scale values of the recorder and range is the region covered by the span by specifying both end-scale values. If we speak of a one millivolt recorder, we are actually saying that the span of the recorder is one millivolt, from zero to full scale. Many recorders will also accept and balance against a signal below zero, so a more precise statement of recorder operation would give the range also.

Dead band is defined as the range through which the measured quantity (the input) can vary without causing response. It is commonly expressed in per cent of full-scale deflection. Determination of the dead band of a recorder is a fairly easy operation. A D.C. signal (usually with a millivolt box) is connected to the input terminals of the recorder so that the pen is balanced some place off of zero. The pen is now physically forced a few per cent away from this balance point by turning the motor pin on gear. Now release the gear and allow the pen to rebalance. Then force the pen an equal amount in the opposite direction, release, and allow to rebalance. The difference between the two rebalancing points, expressed as a per cent of full scale, is the dead band. A dead band in excess of 0.5% of full-scale will cause faulty recorder operation, especially missing of very small peaks in capillary GC operation.

The input impedance is the sum of the resistance and impedance of the input circuit of the recorder; it is of little value to the user. What is more important is the source impedance. Source impedance is the resistance the recorder "looks at" in the attenuator output, and this must be in the range specified for a given recorder for linear operation.

Pen step response speed is defined as the time it takes for the pen to come to rest in a new position after an abrupt change to a new constant value in the input signal. If the recorder is balanced at zero and suddenly a one millivolt signal is imposed on the recorder, the time between the application of the signal and the time the pen comes to rest at one millivolt (e.g., full scale for a one-millivolt recorder) is the response time or pen speed of the recorder. A value of one second is common. Slewing and damping are the terms used to specify the deceleration of the pen. Since the chart pen represents a dynamic system in search of new rest point, overshoot of the rest point will occur unless some means is provided to decelerate the pen as the balance point is approached. If the pen is over-damped, the pen will only gradually approach the final balance point. Many recorders have an RC-circuit in the system which provides a means to anticipate the final balance point and cause deceleration before this point. The damping adjustment controls the variable resistance in this circuit.

Interference rejection is a measure of the recorder's ability to operate without change of dead-band or calibration in the presence of extraneous A.C. signals. An alternating current has the same effect on the input terminals of the recorder as it would on the input signal to the electrometer, that is it would create a current flow because of the changing magnetic field associated with the A.C. field (generator effect). The main source of A.C. interference is from the A.C. power lines into the instrument so that the stray signal will have the same frequency as the power line (60 c/s/). Most commercially available instruments have filtering circuits and shielded inputs to eliminate or reject extraneous signals. These provide a barrier to the higher frequency stray signal and paths through which the signal can flow to the ground rather than to the recorder.

The gain or sensitivity adjustment of the amplifier is needed to reduce the dead band of the recorder. Since the measured signal is in the millivolt range, the error signal (input signal minus reference signal) will be in the microvolt range and a large amplification of the error signal is necessary. To adjust the gain, turn the gain adjustment control until the pen starts to oscillate, then reduce the sensitivity slightly (turn the control in the reverse direction).

The damping adjustment is a means of varying a resistance in an RC-circuit as previously mentioned. The damping adjustment normally is included in the measuring circuit panel. To adjust the damping requires application of a fractional millivolt signal. The response of the pen is observed upon application of a 10-50% span signal. If the pen undershoots (is sluggish), the system is overdamped or if the pen overshoots, the system is underdamped. Trial and error are needed to bring the pen to the final rest point with the least over- or under-shoot in the minimum time.

Slide-wire cleaning should be a routine, perhaps weekly item. Naphtha, or other solvents leaving no residue (such as methanol) can be used. A stiff brush should occasionally be used to remove deposits which accumulate between the convolutions of the slidewire. At this time inspect the slidewire for signs of excessive wear, especially in one spot. Since there are many different types of writing mechanisms, pen cleaning may take several forms. It may only require replacement of a ball point pen or replacement of a solid ink supply used in some recorders. The best bet on this point is to consult the recorder manual.

RECORDER PROBLEMS AND SOLUTIONS

No response can mean that the recorder is not plugged in (simplest case), the power is not turned on, the input leads are not properly connected, or, in some cases, the chart drive is turned off, as in some instruments the chart drive switch also inactivates the measuring circuit. Mechanical binding in the balance motor or the pen carriage can also prevent response. Check the servo motor to make sure it is free to turn. No response or sluggish response may be due to a faulty amplifier. Check the amplifier tubes.

If the instrument is driving against a stop full-scale, check the output of the constant voltage circuit, the measuring circuit, or the amplifier according to the manual. This may also be an indication of improper shielding which is allowing a stray signal to be impressed on the input terminals. Check for continuity of shielding. Also check the incoming signal from the electrometer or TC (thermal-conductivity) cell to see whether some large voltage output may be overdriving the recorder.

Sluggish response may be due to the gain adjustment set too low or overlapping. Check these points as listed above.

Erratic or jerky pen movement is an indication of a dirty or uneven slide-wire, a worn contact, mechanical binding, or under damping.

Non-linear attenuation of signal was discussed previously. Check the recorder zero in the absence of an input signal. Turn the attenuator to infinity or remove input leads.

Pen oscillation indicates gain too high, or damping too low. A poor ground connection or a loose input connection will also cause pen chatter, as will a worn slidewire. Check to see if the oscillation or chatter is dependent on the position of the pen along the scale. If so, check for a worn spot on the slidewire at the position of noise.

Failure of the pen to return to baseline after a peak is due to a large dead band if associated with a recorder problem.

Erratic chart drive is due to slippage in the chart drive clutch or slippage of the gears.

If the pen searches all over the chart without finding the recorder zero or over-responds to chromatographic signals, this usually is associated with failure of the battery in the recorder reference circuit, and the battery (usually a mercury constant-voltage cell) must be replaced.

The recorder manual provided with the instrument will normally contain a section describing the common symptoms and remedies. Thus, the last point to be made should probably have been the first - READ THE MANUAL. A recorder checker mechanism is described in Appendix E.

INTEGRATION TECHNIQUES

Quantitative analysis of the chromatogram requires some quantity of component being detected to be measured. The simplest quantity to measure is the peak height since this can be read directly from the chart paper divisions or measured with a scale. In many applications peak height measurements are sufficient. The technique does, however, require precise calibration of the instrument by injection of varied amounts of the component(s) of interest under identical conditions. If the analytical conditions are maintained constant, the peak height will be a direct measure of the component within the linearity of the detector. A quantity more amenable to automatic measurement and better quantitation is the peak area. The triangulation method (H x W 1/2) is time consuming and is subject to human error. The more measurements that have to be made, the greater is the chance to make a mistake.

The technique of cutting out and weighing each peak suffers from the same disadvantages; time consuming, subject to error, dependence on accuracy of cutting, homogeneity of paper, accuracy of weighing, etc. All that is required is a pair of scissors, a lot of patience, and an $800 analytical balance. For this price or less you can buy a disc integrator.

Similar arguments can be used against the planimeter. From an initial investment standpoint, it is cheap, but from an operational standpoint it is quite expensive since it is time consuming and fraught with sources of error (mainly human).

The operation of the disc integrator has been fully described in the literature (4) and on manufacturers' brochures. One of the features of the disc integrator which is highly objectionable is that it is dependent on the recorder operation and any trouble in the recorder will also affect the accuracy of the integration. Change of baseline is not automatically compensated for and a correction must be made between peaks (zero count rate). Each peak must be attenuated to a peak height less than full scale, which may require a trial run on each sample. For the analysis of fairly simple samples with stable analytical conditions (no baseline drift or shift) and a predictable composition, the disc integrator is a good investment. Little can actually go wrong with a disc-type integrator as long as the mechanical connections to the recorder measuring pen are intact and functional. This should be checked daily.

The Daystrom integrator was marketed in this country, but is no longer available. It is included only to point out the principle of integration. Integration was based on the charge (Q) developed on a capacitor with applied voltage (V) ($Q = C \times V$, where C is the capacitance) and was essentially automatic. The signal was automatically attenuated and the capacitor discharged when the peak reached full scale. The capacitor discharge was indicated by a pipping pen and the residual charge on the capacitor at the termination of a peak was recorded as a defection of the recorder pen as some per cent of full scale. The area of the peak was the sum of the pip plus the size of the final deflection. Thus, as with previous methods, line counting was required to translate the integrator output into a number proportional to peak area. No maintenance or troubleshooting hints will be discussed for this type of integrator.

Digital integrators now provide completely unattended integration of the chromatogram. The aim is to provide the analyst with a digital read-out which is a measure of the peak area. The only errors will be in the ability of the analyst to relate this number to the composition of the sample and the fidelity of the detector-electrometer-integrator combination to produce a number which is truly indicative of the concentration. The integrator will produce a number which varies as the size of the electrical input that it sees. The fact that the integrator gives a printed number does not give gas chromatography any more accuracy than the analyst is able to build into his analytical scheme. The fact that this number may be larger than the output of other integrators likewise does not improve the accuracy of the technique.

For complete automatic operation the integrator must have certain features. There are several instruments on the market today which include all or part of these features. One required feature is a high count rate, expressed in counts per second per millivolt of input. Coupled with the count rate must be the ability to print out a large number of counts, since a high count rate may produce an output in the tens or hundreds of thousands of counts for large peaks. A high count rate is also essential so that good statistical reproducibility can be obtained on small peaks. The linear dynamic range of the integrator is the ratio of the largest to the smallest peak which the instrument can handle within the system's linear region of response. If the integrator can give a linear response to a 50 mV signal as well as a 10 microvolt signal (0.01 mV), the linear dynamic range (L.D.R.) is 50/0.01 or 5,000. Extension of the L.D.R. is accomplished by auto-ranging: as the signal approaches a certain preset level, the instrument automatically reduces the amplifier sensitivity down by a factor of ten and also automatically feeds the output signal to the next higher decade range of the counter.

The ability of the integrator to handle a variety of input signals encompasses many variants, some of which are shown in Figure 12-5.

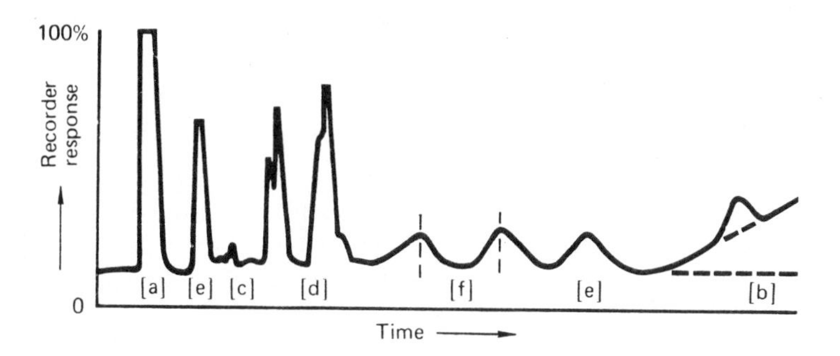

Fig. 12-5. Various chromatogram situations encountered by digital integrators (10).

The input signals may range from a few microvolts to a few hundred millivolts (a). Baseline instability caused by drift must be recognized as such and compensated for by the integrator (b). Noise, usually higher-frequency noise, must be rejected by the integrator (c). Incomplete resolution of components is a common occurrence in GC and presents another challenge to the integrator (d). The incompleteness of resolution may manifest itself as peaks of the same or vastly different sizes or peaks with or without a defined valley between them (shoulders).

The first peaks eluting from the column are normally narrow and quite sharp, whereas later peaks may rise rather gradually and be many seconds or even minutes wide, depending on the analytical conditions (e). With capillary columns the earlier peaks may be only a few seconds wide, while the last peak may be 30-60 seconds or greater in width. In addition to a varied peak width certain peaks may have a definite asymmetry, with either a sloping front or back (f).

Superimposed on the D.C. detector signal will be some A.C. signal from various sources. The integrator must have some means whereby we can control the response to these A.C. signals or a filtering network to remove them.

Basically the logic of an automatic integrator uses three parameters for detection of peak on-set: the slope of the signal (rate of change of millivoltage per second), the amplitude or size of the signal, and the frequency of the signal. If the signal is recognized as the on-set of a peak (see (Figure 12-6), the amplified signal is sent to a voltage-to-frequency converter which has a frequency output that is in direct proportion to the output of the amplifier.

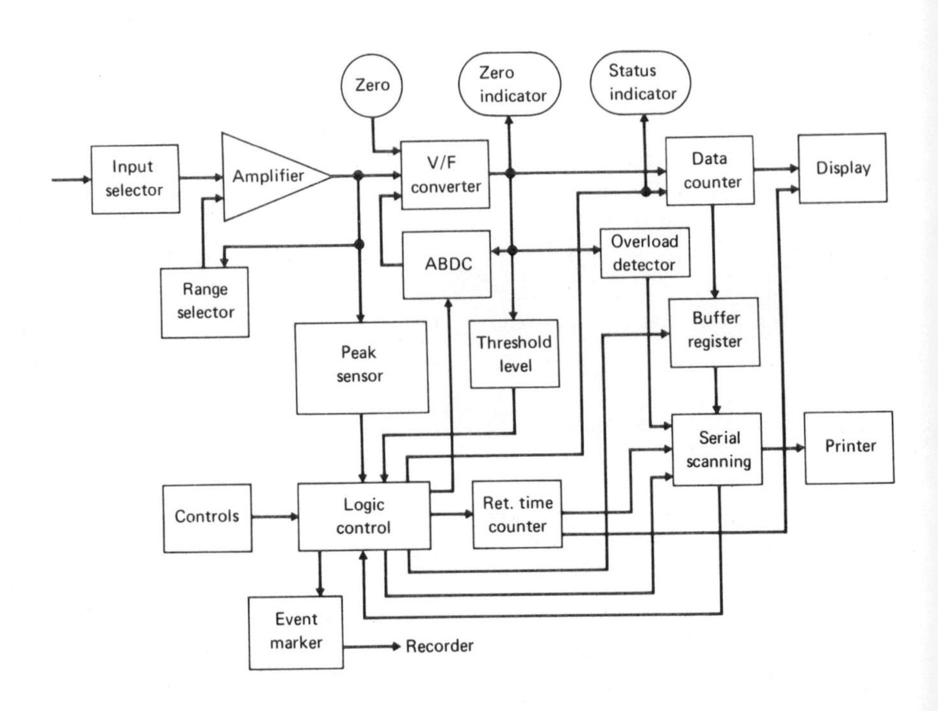

Fig. 12-6. Block diagram of automatic digital integrator (Courtesy of Infotronics Corp.).

The frequency pulses are set to the appropriate decade of the data counter, and a visual display of the accumulation may be provided. The peak sensor logic also includes provision for signaling the end of a peak at which time the accumulated count in the data counter is printed out in digital form. For identification purposes circuitry may also be provided to detect peak maximum (slope goes from positive to zero to negative) and store in a memory system the time which is printed out along with the data count. The read-out consists of a retention time count (up to 999 seconds from start) plus 5-8 digits of area data.

Slope detection works on the change of input signal from zero to positive (increasing signal) to detect on-set and change from negative to zero slope for peak end. This eliminates an absolute reference level and allows integration of overlapping peaks. Provision may also be made for integration when the slope goes from positive to zero to positive (front shoulder) or negative to zero to negative (back-side shoulder). Frequency filtering is used to eliminate high frequency noise or spikes. Amplitude or trip level adjusts the level to which the signal must rise before being recognized as a peak. The value of the level control is the percent of full scale to which the signal must rise before integration commences. For low noise level systems, a low value of 1-3 is used. This insures small signals will be integrated and increases the slope sensitivity.

Correction for baseline drift may also be included. This device works in conjunction with the slope and amplitude controls. If drift above the trip level occurs at a slope less than that required for the logic to recognize it as a peak, the drift correction automatically raises the reference level or baseline. It does not affect the output to the recorder so the drift still appears on the chromatogram. This feature is inoperative during the elution of a peak and hence a certain error will exist if the drift is severe and the peak is fairly broad. At the end of the peak the drift corrector will again commence operation and the base level will be brought to the level existing at that time. This feature is of no value for closely spaced peaks on a continual drift since it will not work except during the time between peaks (minimum of 2-5 seconds needed to sense baseline).

There are also options available known as exponential decay correctors; for accurate integration of peaks occurring on a severely tailing peak, for example, from solvent.

CHARACTERISTICS OF COMMERCIAL INTEGRATORS

A typical digital integrator has a solid state v/f converter input system which has slope, amplitude and frequency peak-sensor logic. A slope as small as 0.01 μv/second can be sensed as a peak. The majority have a linear dynamic range of 500,000, which is certainly adequate for most situations.

Typical count rates of about 2,000 c/s/mv with a maximum count-rate of about 10^6 c/s with a capacity for 8 digits are common (00,000,000). A three digit retention-time readout is standard. Most units can handle two peaks/second if necessary. An auto-ranging device allows inputs up to 500 mV, with auto-ranging occurring at 50 mV. Such integrators can easily handle all commonly encountered GC input signals. Many output options are available including paper tape, magnetic tape, card-punch, and computer.

Today's digital integrators are available in a variety of models and are extremely complicated pieces of electrical equipment. Actual troubleshooting by the laboratory analyst is difficult other than the correct setup of function controls as described in the user's manual. However, certain routine maintenance procedures can be followed to keep the integrator operating efficiently. Unless the integrator is not going to be used for long periods of time (longer than one month), do not turn the electronics off, leave the unit in the reset position. This helps keep the internal electrical components at a constant temperature, and thus helps prevent failure of components due to temperature changes. The ventilating-fan area should be kept clear of accumulated dust and dirt, as this can cause undue temperature build-up in the cabinet, and result in thermal failure of solid-state and/or tube-type components. The only integrator failure one of the authors has had with a digital integrator in six years of operation was due to a build-up of dirt and poor air circulation within the cabinet, which resulted in a few dirty contacts that were rapidly cleaned and resoldered. Also, be sure the integrator is placed in an area that allows good air circulation. Removal of the top of the cabinet or sliding out the equipment chassis (depending on design) will allow dirt and dust removal from the interior of the unit by directing a fine stream of air on the dirty areas and prevent malfunctions in these areas due to dust accumulation. Other repairs, except for checking and changing fuses and function lights when required, should best be left to the experienced service engineer.

COMPUTER LINKUPS

In this section only the equipment available will be discussed, as troubleshooting these systems is for highly trained technicians, not the average chromatographer (including the authors). Very few analytical laboratories have the facility for on-line computer handling of GC analytical data due to the huge capital expenditure required to purchase the computer components. Only the very largest of laboratories and modern manufacturing plants have such a capability. Also, there are many laboratories in which the routine GC analysis load is not sufficient to justify even a minimal on-line system. However, there are workers involved in non-routine chromatographic research who could benefit greatly from computer automation, but for a variety of reasons (mainly financial) cannot go on-line with a computer having sufficient computing power and/or data-storage capacity to meet their particular needs. In such situations as these, off-line data acquisition and control systems can be utilized to permit the gas chromatographer to realize the benefits of computer data handling at a cost he can afford. As a matter of definition, the *on-line* computer is interfaced directly to the GC unit, while the *off-line* system produces a recorded digital output which is submitted for computer processing at a later time (5). The digital output data to be stored is usually produced via magnetic or paper-punched tape, or a card-punch. In this section we will briefly discuss off-line systems, on-line time sharing systems, small dedicated computer systems, and in-house on-line computer systems.

The primary feature which characterizes the off-line computer system is the indirect link between the digital output of the unit and the computer. A second feature is the lack of a direct control, i.e., from the computer to the GC system and the presence of a system control module. Off-line systems are further characterized by modularity. The chromatographer can choose from a variety of data acquisition systems as well as from numerous computer facilities. The system usually consists of a dedicated coupler (i.e., specially built interfaces to measuring instruments) and recording devices such as paper-tape punch, magnetic tape, or IBM card punch. After the data are recorded, the tape or cards are taken to the computer for processing and printout. One of the main advantages of this method of data acquisition is that the tape and/or cards provide a permanent record of the data, as well as being quite simple to operate. The recorded data may then be taken to a large or dedicated computer for processing or, if paper-tape is the data gathering medium, to a time-shared computer terminal.

For routine analyses, the analog to digital convertor most widely applicable to off-line chromatographic systems is the digital integrator. As discussed previously in this chapter, such units are available commercially with interface circuitry installed to allow direct installation with various off-line data acquisition systems as well as on-line with computers. If a time-sharing computer is available in the laboratory, a common teletype facility with keyboard printout and simultaneous paper-tape punch can be used. The retention time and peak area data are printed out, as well as being punched on paper-tape. The data on the paper-tape is then fed into the time-share computer. The only drawback with this system is that the teletype terminals yield slow data-print-out (10 characters/second), and for some rather involved programs for calculating solution properties from compositional data, it would take as long as two-three hours to print out the data. This could be expensive as well as nerve wracking. However, for most routine-type analyses, this is an ideal off-line system. Another often used system for data acquisition is magnetic tape. This is a particularly good system for routine analyses from which detailed calculations are required, as most large computers have a magnetic tape data input system. The last system for off-line data acquisition to be discussed is the IBM card punch. These are easily interfaced with commercial digital integrators and result in a data card deck for each analysis, with an individual data card for each peak in the chromatogram. Such a system is invaluable in the analysis of a complicated mixture such as motor gasoline, as each card (representing a peak) may be identified by a code number to obtain the detailed composition of a gasoline as described by Sanders and Maynard (6). To process a data card deck, a large computer is usually required, and such a system is only really useful when detailed calculations are required that need a great number of characters printed-out. The other previously mentioned data acquisition systems are more practicable for routine analyses.

With the advent of time-sharing computer access being available in almost any area where long-distance telephone service is available, the facility for on-line usage of a remote computer is possible. Dialing the computer's phone number puts the user in direct communication with the computer, and he is charged on the basis of computer time used (usually $5-$20/hour + fees). The program data are permanently stored in the computer and are recalled in a fraction of a second to do the desired calculations. The interface between the computer and the digital integrator is usually the teletype. The advantages of such a system are conversational data reduction of on-line recording, permanent record and printout of results, and simple type of operation. Also, the soft-ware required (programs) for most calculations are quite simple, either written in FORTRAN or BASIC (a conversational language, easily learned by someone with no previous computer experience). The major drawback is that only one chromatograph can be connected to a teletype terminal at one time (although several can be connected simultaneously to the remote computer with separate teletype units) and the type of analyses handled would have to generally be routine that the computer has been previously programmed to handle, with peak identification by elution time or by a retention time index technique. Such a system would not be very useful for quantitative analyses of non-routine samples, but would be quite useful in non-routine analyses where a sophisticated computer is required for complicated chromatographic interpretations (for example, the evaluation of physiochemical parameters from measurements of central moments of the chromatographic peak profile) (7).

Dedicated computer systems generally use small computers (although IBM 1800 systems are common) that process data in "real time" (simultaneous with the analysis) providing information immediately for the researcher in the laboratory. Such units are dedicated to process only the type of data they are programmed for (hence the name), and must be reprogrammed for any different type of analysis (as does any computer). However, due to their small capacity, dedicated computers can only handle a limited number of analyses. The main advantages to this type of computer are high recording speeds (to 10kHz characters/second), immediate display of results, and on-site control of the GC unit parameters, such as column oven control, sample injection, etc. The main disadvantages are: (1) programming skills are required to operate effectively, (2) generally no permanent input data record is available and, (3) the systems can be very complex and expensive. Thus, unless built especially for handling GC data (and there are several such units now commercially available), the small dedicated computer would be difficult for the average chromatographer to handle. However, units built specifically for monitoring GC digital integrator outputs, and not for controlling analysis parameters, are easy to operate and especially attractive for the GC expert who can spare 10K - 50K dollars for the unit.

Within the past few years, minicomputer systems have been introduced that have off-the-shelf software which allows the rapid processing of data by all of the standard GC techniques (including internal standard calculations, area % calculations and weight % calculations using detector response factors), as well as peak identification of up to 250 named peaks/analysis. Such data systems also have logic control for automatic liquid sampling systems, superior software for compensation of baseline drift, and the ability to handle up to 16 chromatographs simultaneously. Such systems normally required from 16K to 32K words of core, especially if simulated distillations and capillary-GC analyses are being done simultaneously, as these programs make use of large segments of core. These are also digital computers, such that the analog to digital converter is placed next to the chromatograph, allowing the digital signals from the A/D converter to be sent up to 1000-feet to the computer, if necessary, without picking up any interfering noise along the transmission line. Such minicomputer systems are not only valuable in the research laboratory, but also are very valuable in routine quality control operations. As far as troubleshooting such systems, it is beyond the scope of this book (and the authors) to discuss attempted repair work on such sophisticated solid-state systems. The best advice that can be given with respect to the complicated minicomputer-based data systems of today is if it isn't working correctly, call the computer field engineer for your particular unit.

One question often asked about GC-minicomputer data systems is with regard to a noisy chromatograph being linked to a sensitive computer, and will it cause untold operational headaches. Like all modern digital integrators, digital computer systems have routines built into the software package to recognize noise spikes within limits as specified by the computer software. The computer system will ignore noise spikes as long as they are faster than the minimum peak time specified in the software. Minicomputer noise rejection logic is somewhat the same as that used for digital integrators, only more precise and effective.

The in-house, on-line large computer system is that which is uselessly visualized by many chromatographers, especially if their lab is small, because of the prohibitive cost of large computers. Although such computer linkups are of great value, they are not a panacea to the chromatographer's data-handling woes. Even though instant results are achieved at the end of an analysis, and the computer can be programmed to control the GC analysis parameters, the analytical data are not recorded, and often an analysis must be repeated only because the results "appear" to be in error. This type of linkup should probably produce better quantitative results than the digital integrator, as a much better job can be done in correcting peak areas of partially resolved peaks and proportionally correcting for a drifting baseline. However, a good knowledge of programming is required even to begin in this area (or immediate access to a Systems Analyst), and still the computer could only positively identify peaks with regard to peak patterns, elution times, and retention indices. Such software techniques can be used for isothermal and programmed-temperature runs of relatively simple mixtures; however, for extremely complex wide-boiling mixtures (such as gasolines) that require broad temperature programs which are difficult to repeat exactly, the computer can be fooled and peaks misidentified. Regardless, at some time in the future, the on-line computer will probably replace completely the off-line systems, and it is important that all of us with responsibility in GC analyses remain aware of developments in on-line techniques. However, in the near future, off-line data acquisition remains the most attractive means of automatic data handling for the majority of laboratories. While on-line systems have the potential ability to produce better analytical data than the digital integrator, this has not been conclusively borne out by experimental evidence (8), and the experience we gain in using off-line systems should be quite valuable when the great transition occurs.

Before concluding this chapter, in which we have discussed techniques used to measure the output of GC detectors, we should first consider the broad topic of quantitative analysis by GC. Quantitation in gas chromatography may sound like a curious topic to be discussed under chromatographic maintenance and troubleshooting, but with a little reflection it is easy to realize that many of the problems which occur in gas chromatography, which ultimately involve troubleshooting, begin when incorrect answers are being obtained for an analysis. Very often the problem is traced to leaks in septum caps, or problems of column contamination, etc. But, sometimes the problem is merely a matter of how the quantitative analysis is being carried out; and, therefore, an understanding of the available techniques to quantitate gas chromatographic analysis and, in particular, a knowledge of the pitfalls of some of the quantitative techniques are needed in order to be able to say definitely that there is a problem with the gas chromatograph and that troubleshooting procedures must be started. Surprisingly, many of the problems in quantitation do not come with the calculation technique being used in gas chromatography, but come even earlier, during the measurement of peak areas. One of the original techniques for measuring peak areas is the cut and weigh technique, where the peak area is cut out of the recorder chart and the paper weighed. This method is probably not used very often anymore. The manual technique is measuring with a ruler the peak width at half-height and multiplying this 1/2-width times the height of the peak, otherwise called triangulation. Then there are the more mechanical methods for peak area measurement such as the disc integrator, but this still requires considerable manual work in counting peak areas. At present, the automatic digital electronic integrator is most often used to monitor GC analyses. Also, on-line computers are being used with increasing popularity, as just discussed in the preceding paragraphs. However, in most cases, the area integration is very similar to that used with an electronic integrator.

Turning first to some of the problems with measuring area by triangulation: Consider two peaks--the first is a gaussian peak and the second is a non-gaussian peak, such as a typical tailing peak from a gas chromatograph. Let us consider what the area measured by triangulation is and how does it relate to the true area of the peaks. The height of the peak we will call H. The width at half the height equals

WH. Therefore, the area of the triangle approximately equals H x WH. This does not equal the actual area of the peak; and, for a gaussian peak, the difference can be calculated. The relationship of the area of a triangle thus calculated is equal to 0.94 x the area of the actual peak. With a non-gaussian peak, it is impossible to specify a relationship between the true peak and the area of the peak obtained by triangulation, particularly if the chromatogram is composed of gaussian and non-gaussian (or tailing) peaks. This is only one source of error in triangulation. There is an even worse situation if the chromatogram also contains partially overlapped peaks. When comparing triangulation with electronic integration, the situation becomes even more difficult, because the electronic integrator will measure the peak area by dropping a vertical from the valley to the baseline. In this case, the peak area assigned to the two peaks is totally different from the areas obtained by triangulation, and can, in fact, reverse the order of the peak area sizes.

It is obvious from this first brief discussion that the way in which the area of the peak is calculated has a considerable effect on the accuracy with which it can be quantitated. It is also quite possible that differences in analyses of chromatograms, which contain partially resolved peaks or contain tailing peaks, and the differences in the analyses between two workers, can be entirely related to the way in which the areas are calculated, i.e., whether they are calculated by triangulation or by electronic integration.

Turning now to the basic calculation techniques used in quantitative gas chromatography, let us discuss what is involved, or what is assumed, in using the different techniques. First of all, normalization. In this case, the area of the peaks are divided by their response factors, and these resultant values are summed to get the total weight-corrected peak area. Then the corrected area of each peak divided by the total weight-corrected peak area multiplied times 100 will give a weight percentage composition. In this method there are two essential requirements. First, all components must be detected, measured, and identified, and second, accurate response factors must be known for all components. These are minimum requirements for approaching reasonable quantitation. Obviously, the simpler technique of just straight area normalization without using response factors is even more inaccurate and should not be used if any other technique is available.

Using the corrected area normalization technique, one is dependent on having accurate response factors for all components. Normally, when using a thermal conductivity detector, response factors can be taken from the literature, although there are still two cautions to be observed.

First, one must make sure that the detector is not overloaded for the major components, so that the thermal conductivity detector is still operating in its linear range. Second, when doing the analysis of certain inorganic gases, one must be very careful because the thermal conductivity of the mixture with helium is not linear. For instance, hydrogen. When working with a flame ionization detector, it is doubtful that response factors are transferable from one detector to another with any level of confidence. However, they can usually be applied to an accuracy of about 10%, provided that they were determined on a detector where the flame conditions are optimized and are used with a flame detector where the flame has been optimized. When a flame ionization detector is operated out of its optimum range, the response factor of components can change quite seriously.

A somewhat better technique for quantitation in gas chromatography is comparison with a standard mixture. In this case, the component composition is determined by ratioing the peak areas of the unknown material with the area obtained for a standard material. Essential requirements for this method are that the components have been correctly identified, the standard sample analyzed to the required accuracy is available, and the sample size used with the standard mixture and the sample is exactly the same. This technique is very frequently used in process chromatographs and is an excellent technique for use when the measurement of only one or two components in the mixture are required.

The best and the most accurate quantitation technique in gas chromatography is the method usually called internal standardization. In this case, a known material of known concentration is added to the sample and the peak area of the required components compared to the peak area of the standard. By a graphical technique, the concentration of any component may be plotted against the ratio of the peak areas of the component and the internal standard. This is used as a graphical technique in order to point out that the method is also applicable when a detector is in its nominal operating range, although very obviously it is preferable to work in the linear range. The use of a single response factor for an internal standardization technique assumes that one is working within the linear range of the detector; and, as with other techniques, this should be checked to make sure that the detector is linear by running standards which contain a greater concentration of the required material than are normally found in the samples. The essential requirements of this technique are to be sure that care is taken in choosing the internal standard, and that the concentration of the internal standard must be the same as used when obtaining information for the calibration curve. In addition, the calibration standards must be accurate. The internal standard should be chosen from material similar to the components being analyzed. For instance, if the analysis is for a hydrocarbon, then a normal parrafin is a suitable internal standard. If, on the other hand alcohols or esters are being analyzed, then it is preferable to use a standard of the same chemical type. The internal standard must not occur naturally in the sample being analyzed, and it must be clearly separated from all of the other components in the mixture by the chromatographic analysis. Given these requirements, the internal standardization technique can be very accurate and give chromatographic results with a relative error of less than 1%.

In summary, when doing quantitative work in gas chromatography and suspecting that there is a chromatographic problem, it is best to insure that the problem is not merely associated with the type of quantitation used for a perfectly adequate chromatographic separation. This is very often an overlooked place for errors. The error in the quantitation can arise from the method by which the peak areas are measured. It can certainly arise from using area normalization techniques. In this case, very large errors may be obtained. The best technique is that of using an internal standard. Further, using a computer to measure peak areas is no guarantee of correct results. Using a computer with software that employs a poor area integration routine is an excellent way of getting an incorrect answer to four decimal places.

In this chapter we have discussed the function and operation of various components in the electrometer/recorder/digital integrator (or GC data producing) system, and described some routine maintenance and troubleshooting techniques. The interfacing of GC data producing systems with computers was also discussed. Due to the complexity of these interfacing arrangements, no maintenance or troubleshooting procedures were discussed, as this could be the subject of another entire book, not just a small part of one chapter. The analyst can often troubleshoot and repair electrometers and recorders (with careful reference to the users' manual); however, troubleshooting of complex solid-state digital integrators should be left to experienced service engineers.

REFERENCES

1. L. S. Ettre, *Open Tubular Columns in Gas Chromatography*, 1st Ed., Plenum Press, New York, 1965 (p. 133).

2. Hewlett-Packard Detector and Controller Manual, 1970, Hewlett-Packard Corporation, Avondale, Pennsylvania.

3. Varian-Aerograph Model 1520 Users Manual, 1969, Varian-Aerograph Corporation, Walnut Creek, California.

4. S. DalNogare and R. S. Juvet, *Gas-Liquid Chromatography-Theory and Practice*, 3rd Ed., Interscience Publishers, New York, 1965, p. 255.

5. J. G. Peddie, "Research and Development," April, 1971 (p. 26).

6. W. N. Sanders and J. B. Maynard, *Anal. Chem., 40*, 527 (1968).

7. O. Grubner, *Advances in Chromatography*, Vol. 6, New York (1968).

8. J. E. Oberholtzer, *Journal of Chromatographic Science*, Vol. 7, December, 1969 (p. 720).

9. J. B. Maynard, Private Communication, 1971.

10. M. T. Jackson, Private Communication, 1971.

QUESTIONS AND ANSWERS

Question: What is the best way to clean dirty electrical contacts?

Answer: One of the best ways to clean electrical contacts of the flat-wafer variety (like found in a GC recorder-output attenuator, for example) is with a clean and fairly firm pencil eraser. For dirty contacts in hard-to-reach places, there are several aerosol cleaners now available on the market. They require simply spraying the contacts and then working the switch back and forth several times to allow the solvent in the aerosol spray to clean the scum of dirt off of the switch contacts.

Question: In a GC system equipped with a glass column system and a flame photometric detector (FPD), a baseline is observed as shown below. What could be causing this problem of baseline stepping?

Answer: The "baseline stepping" shown above could come from electrical noise, such as the pickup of an AC current, as would occur from the cycling of an oven heater, or some other off-on AC device. Another possible cause could be some cyclic change in the gas flows to the detector, such as a particle that is intermittently restricting flow in a gas supply line. The cure could be to better electrically insulate the cables running from the FPD to the electrometer to eliminate pickup of AC currents in the vicinity of the cabling. Another possible source could be cyclic noise emitting from the photomultiplier tube power supply. There could also be a "ground-loop" through the detection system wherein AC currents could be picked up. The answer here is to isolate the ground for the detection system from the remainder of the unit. In addition, all the gas flows should be checked at the detector to determine whether there is any intermittent loss or restriction of any of the gas flows.

Question: What is the best quantitation technique for overlapping peaks?

Answer: In the case of overlapping peaks, peak triangulation is a fairly good technique for estimating the concentration of each component. Probably the worst system to use would be a ball and disc type integrator. Most modern integrators would do an adequate job of estimating the amount of each component, even if one of the peaks is only a shoulder on the other peak. This is usually done by dropping an imaginary line from the intersection point to the baseline. Some GC minicomputer systems have software packages that allow more accurate calculation of the components by storing bit information for the peaks until the baseline is obtained again, and then calculating the estimated contribution of one peak to the other. A discussion in quantitation is presented at the end of this chapter.

Question: What kinds of events contribute to the variability of peak height (area) ratios when using an internal standard?

Answer: The choice of the internal standard is important. It should have exactly the same (or as close as possible) response characteristics as the samples to be analyzed. If this technique is followed, the major variable could then be the size of sample injected, and the ratio of the amount of internal standard to the material being measured. Most detectors exhibit varying responses to the amount of sample being detected, especially if the internal standard component is several orders of magnitude more concentrated than the components being measured. Sample size should be kept constant, and small enough so as not to saturate the detector or attendant integral electronics.

Question: What could be wrong if the peak area (integrator counts) does not double when the attenuator is switched to the next lowest position?

Answer: In most modern-day auto-ranging integrators, the full detector (or amplified detector) signal is sent to the integrator, with the attenuator only active through the strip-chart recorder. However, if the attenuator is connected to the integrator, the problem is one of not having the integrator system at the zero-voltage position when there is no signal (or only background signal) from the detector. Most important, it is dangerous to attenuate a signal during counting of a peak, as the range-change itself could throw off the previously set integrator zero, causing peak area (count) errors. It is best to leave a run on only one attenuation if the attenuated signal goes to the integrator.

Question: Problems exist in zeroing the recorder amplifier, shifting recorder baseline, and background noise. How can these situations be improved?

Answer: The recorder amplifier must be zeroed with the signal attenuator in the "infinity" position which simply shorts the recorder input, allowing zeroing of the recorder electronic package. In most older recorders, if the amplifier cannot then be zeroed, suspect a weak tube, and check them one at a time, replacing any marginally performing tubes. If this doesn't help, check the slidewire to make sure it is clean in the recorder zero area. Beyond this, and with modern solid-state recorders, now is the time to call a service engineer. A drifting baseline can be caused by many other factors than a recorder malfunction. A troubleshooting scheme for detecting the source of the drift is given in Chapter 14. Background noise can also be picked up anywhere within the system, and this can be traced using the troubleshooting schemes in Chapter 14.

Qusstion: A large loss in sensitivity was encountered when using an electron capture detector (ECD). The radioactive foils were cleaned with the prescribed solvents, the carrier gas is well scrubbed (including passage through molecular sieve) and no leaks are detectable throughout the system. The electrometer was checked out, and is operating within manufacturer specifications. What could have caused this gross loss in sensitivity by the ECD?

Answer: This sounds very much like an age problem, especially if heavy samples are continually analyzed that could coat the radioactive foils. If several foil cleanings have been necessary to maintain performance, the loss of some of the impregnated radioactive material, that produces a finite amount of soft beta-radiation, was possibly brought on more quickly by high operating temperatures or repeated contamination. The best approach here is to buy new foils, usually by sending the entire detector to the factory to be refoiled.

Question: What is the best method for screening-out A.C. noise in a GC unit and integrator system located on site in a plant where there are many sources of electrical noise?

Answer: The best approach here would be to power the unit through an isolation transformer, as well as isolated grounding points for the GC-electrometer-detector system and an isolated and separate earth ground for the integrator alone. This would eliminate the chance for electrical ground-loops to become established, thus eliminating integrator noise. In addition, the isolation transformer on the A.C. power line should eliminate A.C. noise from the line being picked up by the detector-electrometer circuitry.

Chapter 13

AUXILIARY SYSTEMS

Since the beginning of separation of organic compounds by GC, there has been a continuing search for simple qualitative techniques to identify the minute amounts of compounds separated by GC other than their emergence time from the column. Many methods have been employed to accomplish this difficult identification task. In this chapter we will discuss several of the more successful auxillary systems used for qualitative monitoring of GC columns as well as some of the problems that are often encountered in their use.

In the late 1950's, a popular method to characterize the components emerging from the GC column was by functional-group wet-chemical analysis. In this technique, the column effluent (usually after passage through a thermal conductivity detector--TCD) is bubbled through a small amount of the desired reagent to determine if a given functional group is present in the peak under scrutiny. For example, an acidic 2,4-dinitrophenyl-hydrazine solution is used to determine whether a peak contains an aldehyde or ketone carbonyl compound. If such a carbonyl is present, the usual 2,4-dinitrophenyl-hydrazone (2,4-DNPH) derivative will be formed as a precipitate in the solution. The precipitate can then be dried, recrystallized (if enough material is present), and a micromelting point determined to help in identification of the peak, as the melting points of 2,4-DNPH derivatives are characteristic for many of the various carbonyls. Another characterizing reagent often used is bromine in solution with carbon tetrachloride, which discolors from the reddish-brown of bromine to a clear solution in the presence of olefinic double bonds. It has been the author's experience that at least 20-30 μg of component be eluted in a given peak to get a positive reaction with most functional-group reagents, although the limit of detection quoted for some of the functional group tests is much lower. To obtain useful results with these systems, 1/4 inch o.d. packed columns are usually employed to allow large enough sample sizes. A detailed description of a useable GC sampling technique, reagents and examples has been reported previously by Walsh and Merritt (1).

The major problems of concern with this type of system are the same as with any other micro wet-chemical system. All gas-bubbling vessels should be kept clean to prevent interference with the primary reaction to be used. Usually a small bore glass or stainless steel extension is attached to the TCD outlet and extended to the bottom of the glass reaction vessel. This tube must be kept very clean, and warm, if possible, to prevent any condensation of heavy components emerging from the chromatograph. The tube should be removed and periodically cleaned with a polar solvent (such as chloroform) to prevent cross contamination between samples examined. If the above mentioned cleanliness suggestions are followed, very little can go wrong with this simple technique for monitoring column effluents. One suggestion is that the very slowest flow rates possible should be used to allow the reagent the maximum time to mix and react with the materials in the column effluent.

The most successful auxillary techniques for qualitative identification of GC peaks have been spectroscopic in nature including infrared (IR), ultra-violet (UV), and mass spectroscopic (MS) techniques. Of the three methods mentioned, the GC/MS technique has been highly refined from the beginning with a Bendix Time-of-Flight mass spectrometer (TOF-MS) monitoring the effluent of a packed column (2) to present day monitoring of capillary column effluents with fast-scanning, high-resolution mass spectrometers. In addition, the methods for interfacing GC/MS systems have been greatly improved, and have been reviewed in some detail by Rees (3).

INFRARED SPECTROSCOPY

Analysis of GC fractions by IR gives considerable functional group and structural information on the fractions. One of the better methods for obtaining IR spectra of GC peaks of components with low vapor pressure is to trap the peak in a cooled (dry ice or liquid nitrogen) flow-through trap, transfer the contents of the trap to an IR cell, and then record the IR spectrum in the usual manner. Another method is to trap GC fractions eluting through a nondestructive detector (such as the TCD) using an ordinary, straight melting-point tube, with about the middle 2-inches of the tube being cooled to dry ice temperature. The same type of trapping can be done from a column split-stream if a destructive detector such as an FID is used. The melting-point tube is usually connected to the GC effluent with a specially machined piece of Teflon tubing that gives a leak-tight seal. After the component has been trapped, the exhaust end of the tube is capped with a silicone rubber septum, the tube disconnected from the Teflon connector, and that end quickly capped with a silicone rubber septum to prevent loss of inert carrier gas and the possible entrance of air and water vapor into the tube. The melting-point tube is then removed from the dry ice and allowed to warm to room temperature. A septum is then removed from one end of the tube, and the open end inserted into a micro-IR cell (NaCl) as shown in Figure 13-1. The sample is then washed from the sides of the tube with a small (1-5 μl) amount of chloroform or carbon tetrachloride into the micro-IR cell sample cavity. If enough

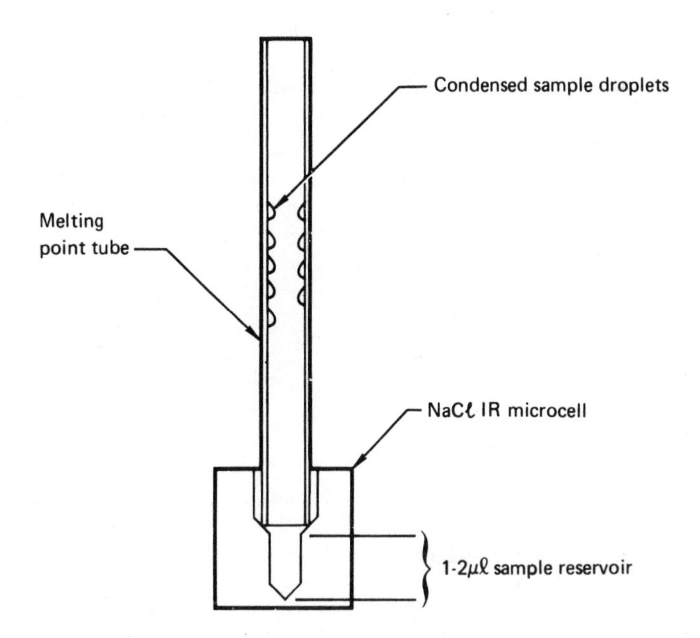

Fig. 13-1. Sample transfer from melting point tube to IR microcell diagram.

sample is present in the trapping tube, the entire assembly may be placed in a small laboratory test tube centrifuge (cushioned in the bottom by tissue or cotton to prevent fracture of the microcell during centrifuging), and the liquid in the tube centrifuged into the IR-microcell sample cavity.

When the sample is in the IR-microcell, a small drop of mercury is placed in the cell above the sample to prevent loss by evaporation and to keep moisture from entering the cell. The surface tension of the Hg is such that it will not drop into the sample reservoir. The cell is then placed in a special microcell holder, and with the aid of a beam condenser in the reference beam, the IR spectrum obtained in the usual way. Commercially available are devices for trapping peaks in KBr, and obtaining their spectra in a KBr pellet; however, such treatment is only useful with very heavy components.

The major pitfalls of using the microcell technique previously described or any other sample cold-trapping method for IR is loss of sample due to inefficient tubes used for trapping. This is especially true if the sample would tend to form an aerosol in the carrier gas at dry ice temperature. Another problem is that of contamination with closely eluting components or even something absorbed from the atmosphere. It is of utmost importance that the cell and trapping tube are washed with reagent grade solvents (chloroform usually used) and dried before use. Also, it is important to cap the tube tightly under the slightly positive carrier gas pressure available at the column exit to minimize the entrance of water and CO_2. Water will not only interfere with the IR spectrum, but will damage the NaCl microcell. CO_2 will interfere with the IR spectrum. To prevent this, the caps should not be removed from the trapping tube until it has essentially reached room temperature. The sample transfer procedure should then be carried out as rapidly as possible. If the microcell technique is used, it is also important that there are no leaks around the Teflon sleeve used to connect the trapping tube to the GC unit outlet. A loose fit will allow back diffusion of air and moisture into the cold trap due to its reduced temperature.

For more volatile components, samples can be taken in evacuated, long-path gas cells. As the peaks are eluted, the column effluent is diverted to the gas cell, and at the termination of the peak the cell is removed and sealed-off by closing the inlet and exit valves. Most gas cells of this type can be heated to about 200°C, if necessary, before recording the IR spectrum. The spectra obtained in this manner (gas phase) show more fine structure because the molecules have more freedom of movement in the carrier gas matrix than in the liquid phase. A typical gas cell is shown in Figure 13-2.

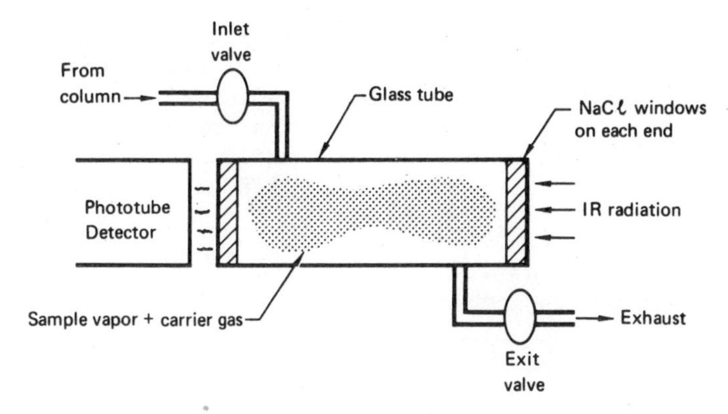

Fig. 13-2. Diagram of a typical gas phase IR cell.

The major problems associated with collecting gas samples for IR analyses occur with loss by leakage, condensation on cell walls, interfering gases (such as back diffusion of water, CO_2, etc. from the atmosphere) and contanimation of the peak being examined by closely eluting components. Fairly large sample sizes are generally required to obtain a sufficiently concentrated gas sample, and as a result, resolution often suffers.

There are commercial IR units available that are constructed for direct interfacing with a GC. These units work on the gas phase principle, can rapidly scan the region from 1.5 to 15 microns in 6 to 45 seconds, and be ready for another component. The heated cells (250°C) are constantly flushed with carrier gas, then when a component appears whose spectrum is required, the column flow is diverted to the sample cell with the exit valve closed. Immediately on capture of the component the inlet valve is closed, and the IR spectrum recorded. Between successive samples the IR absorption sample cell is flushed with carrier gas by opening both the entrance and exit valves. Infrared spectra of known compounds are obtained in the same manner as the unknown for comparison. Two units of this type available are Beckman's IR-102-GC/IR detector/spectrometer and the Wilks chromatograph-IR spectrograph. If sample components are well enough separated (usually by packed columns or large-bore capillaries), these units, along with retention time data, usually give enough information to identify the component peak, if pure (i.e., not more than one component in the peak).

To assure proper operation of a gas-phase IR system, all connections must be routinely checked with soap solution to find any leaks that could result in sample loss or atmosphere diffusion into the cells. Routine maintenance of the cells is accomplished by a high-temperature bake-out at the maximum cell temperature with a carrier gas flow (or purge gas, bypassing the column) of 100-200 ml/min. This is sufficient to remove most contaminants adsorbed on cell walls. If cell construction permits, a further step in cell cleaning is that of evacuating the gas cell while heating. This approach is especially useful for cells of stainless steel construction.

ULTRAVIOLET SPECTROSCOPY

Gas samples can be taken for UV analysis in the same manner as just described for the IR gas phase units. However, liquid samples are usually diluted in micro-UV cells with spectral grade normal-heptane or cyclo-hexane for UV analyses. Such UV spectral scans are particularly useful if aromatic compounds are present. The major problems with collecting GC peaks for UV analysis are sample loss by handling or vaporization (same as for IR samples), and contamination from outside materials. Also, the presence of oxygen from the atmosphere can greatly mask and alter UV spectra, especially in the far UV region. Thus, special precautions must be taken to keep oxygen out of both the gas phase systems and the liquid phase system as this promotes oxidation of the material to be analyzed. If the presence of oxidants are expected, scan the desired spectral area once, pause for a minute or two, then rescan. If the spectra are changing, oxidants probably are present and the spectra will not be representative of the pure compound. Any oxidation of a molecule is significant in the UV, as the C=O or C-OH groups and other O-containing groups are chromophoric and tend to mask information about the hydrocarbon structure of the molecule.

MASS SPECTROMETRY

Gas and liquid samples are collected in cold traps for MS analysis as described previously for IR and UV analyses, except in MS analyses, the sample must be revaporized, and transferred to the ionization region of the MS without contamination or loss. Such systems are necessary if only slow-scanning MS units are available. Many complaints are registered by the mass spectrometry laboratory that the gas chromatographer has just submitted another trap filled with CO_2 and water, and precautions should be taken to circumvent this difficulty.

A trap filled with unexpected contaminants such as CO_2 and water vapor when taken to the mass spectrometry laboratory either indicates a serious leak in the trapping system, or a leak during vapor transfer to the MS. All traps should be pressure tested to one atmosphere, and all connections checked for leaks with soap solution. Traps should be maintained after every use by a vacuum bakeout at 250°C (especially if stainless steel traps are used). One quick way to vacuum test a trap is to flush at high temperature with a dried stream of prepurified nitrogen. Connect this trap to the MS inlet system used and begin to pump away the nitrogen from the trap. While pumping out the trap, continually take spectra, looking for increases in the concentration of water and CO_2. If no increases in the background level of these atmospheric constituents occur during trap pump-down, it is vacuum tight, and may be used for trapping from GC columns without worry of atmospheric contamination by trap leakage. However, much better and quicker techniques are available to the gas chromatographer who needs MS information for GC peaks to aid in their identification.

The successful interfacing of GC units with fast scanning mass spectrometers has been one of the most important developments in the qualitative identification of GC peaks. Since sample sizes required for good mass spectra are so small (good quality spectra can be obtained from as little as 10^{-9} gms of sample), the effluent of a capillary column can be split 50/50 between an FID and MS and still obtain a good chromatogram along with mass spectral data for peaks of interest. The beginning of tandem GC/MS work was begun about 16-18 years ago when Gohlke published his work at Dow Chemical Company concerned with monitoring a GC column effluent with a Bendix TOF-MS (2). The Bendix TOF unit has been extremely useful with GC units, as it produces 10,000 or 20,000 separate mass spectra/second in the 0-220 unit mass range, depending on the desired frequency of operation. Displayed on an oscilloscope, this appears as the complete spectrum for the eluting component and can be photographed with a Polaroid-type camera for spectral interpretation. Later units were then fitted with gated-type scanning units, that allowed scanning a decade (mass 20-200) in about 3 seconds, fast enough to obtain a useable spectrum from rapidly eluting capillary peaks. Even at the low resolution available with the old TOF units, the mass spectra, along with retention time data, provided a powerful tool for identification of compounds purified by the GC separation. However, several modern techniques for interfacing GC with MS units now allow GC units to be connected directly to fast-scanning high-resolution mass spectrometers that have unit resolution values from m/e of 1 to 1,000 (3).

A GC unit carrier gas at the detector exit normally exits at atmospheric pressure with flow rate from 1-200 ml/min. However, most MS units operate best at pressures of 10^{-4} torr or less. Thus, any method to selectively remove the carrier gas from the sample prior to entrance into the ionization region of the MS is certainly desirable to allow a more concentrated sample to reach the ionization region. Several continuous methods for effecting this carrier gas removal have been used. These "molecular separators" fit into the GC/MS system as shown in Figure 13-3 (4).

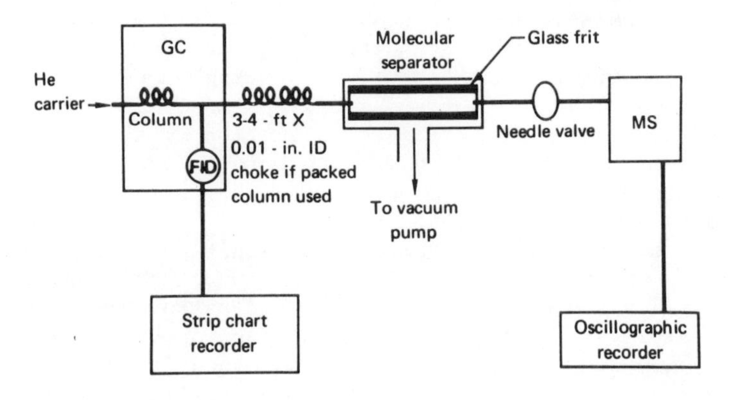

Fig. 13-3. Block diagram of typical GC/MS interfacing.

Some typical terms used to describe the performance of molecular separators are the Yield (Y) and the Enrichment (N). Grayson and Wolf (4) have defined these separator operational parameters as follows:

$$1.\quad Y\ (\text{yield}) = \frac{\text{sample entering spectrometer}}{\text{sample entering separator}} \times 100 = \%\ \text{value}$$

Thus, the % yield represents the ability of the separator to pass sample molecules from the GC carrier gas to the ion source of the MS unit.

$$2.\quad N\ (\text{enrichment}) = \frac{\text{sample/Helium ratio on MS side of separator}}{\text{sample/Helium ratio on GC side of separator}}$$

$$3.\quad N = (Y)\ \frac{\text{Helium flow into separator}}{\text{Helium flow into spectrometer}}$$

The enrichment (N) may be a large and meaningless parameter if the amount of sample reaching the MS is too small for a good spectrum.

In the following paragraphs several different separator systems will be discussed, along with typical maintenance procedures as well as specific problems and solutions when available.

FRITTED GLASS (OR BIEMANN-WATSON) SEPARATOR

This is an all-glass separator introduced by Biemann and Watson (5) (6) and consists essentially of a porous glass tube which acts as a stream splitter, with the faster diffusion of the small carrier gas molecules through the fritted (porous) glass resulting in an enrichment of the compound of interest. Such separators are most effective when operated at flow-rates from 1-50 cm^3/min (making them useful with capillary columns), and at temperatures as high as 350°C to prevent heavy materials from condensing in the fritted glass or on other porous surfaces. The yield (Y) of this separator is in the 30-75% range, and the enrichment (N) in the 10-40% range, depending on GC column flow-rate and the pressure drop across the separator. This separator operates best at flow-rate of 20 ml/min or less.

With very polar materials, some peak distortion is noted with this separator (mainly trailing of the polar components into the MS). Care must be taken when using this separator with unstable organic compounds, as some may react with active sites on the hot glass surfaces. To overcome this difficulty, silanization of the separator with dimethyldichlorosilane has been advocated (3) (usually by injection of about 100 µl of the silanizing agent into the separator when disconnected from the mass spectrometer inlet). Another problem often encountered with these glass frit separators is the adsorption and retention in the glass pores of trace amounts of polar materials. This problem also is most readily solved by silanization of the separator. If low vapor pressure materials are to be examined, the separator and all lines leading to and from it must be well heated. Our experience has shown that even when operated at 250°C, when analyzing 3-4 ring polyaromatics, many of the glass pores eventually become clogged with organic material, resulting in a sharp decrease in the sample yield (Y) to the MS.

MOLECULAR JET (RYHAGE) SEPARATOR

This type of separator, constructed entirely of stainless steel, was first described by Ryhage (7). The column effluent is aligned with an orifice which is optically aligned with another orifice. The carrier gas (generally helium or hydrogen) diffuses away from the line-of-sight more rapidly than the heavier organic molecules, resulting in good enrichment of the carrier gas entering the MS with the organic material. The molecular jet separator is primarily used with packed columns with higher carrier gas flows of 30-50 ml/min, which allowed 40-70% of methyl stearate to be passed into the ion source of the MS (8). The two-stage separator is used in the LKB 9000 GC-MS tandem and numerous published results attest to its reliability. Yields (Y) and enrichments (N) in the range of 40-70% are usually obtained with this type of separator. Maximum operating temperatures are near 350°C, although the Ryhage separator is more generally operated at 200-250°C to minimize thermal rearrangements or cracking of thermally unstable molecules. Peak distortion due to the separator is slight. The main disadvantage of this type of molecular jet separator is that it must be optimized for one column flow-rate, and the design (i.e., a distance between jets) must be changed for use at other flow rates. However, once optimized the separator is rugged and easy to operate, with very little surface interactions with organic molecules.

One of the main items of routine maintenance with this type of molecular separator is to make sure the jet openings do not clog. In general, this can be caused by condensation of high-boiling materials, and this possibility should be taken into account when choosing the operating temperature for the separator. If a jet opening does become plugged, a high-temperature/vacuum bakeout will usually cure the problem by vaporizing the organic plug. However, if the plug is a small particle of some foreign material that was in the carrier gas, usually disassembly and mechanical removal is required. To prevent condensation problems, make sure the GC/separator and separator/MS transfer lines are also well heated, with no cold spots. It is also necessary that all vacuum connections be as leak-free as normally required for MS work. If these few precautions are taken, this design separator should give essentially trouble-free operation.

SILICONE MEMBRANE (LLEWELLYN) SEPARATOR

Llewellyn *et al.* have reported use of a silicone rubber diaphragm as a molecular separator (9). With this separator, the eluted organic compounds diffuse through the diaphragm, which is relatively impermeable to the carrier gas (hydrogen, helium, argon, nitrogen). Both single-stage (one membrane) and dual-stage (two separate membranes) separators have been used. The desired flow-rate range for the single-stage is from 1-20 ml/min and to about 60 ml/min for the dual-stage, with a maximum operating temperature 250°C for either type. A yield (Y) of 40-90% and enrichment (N) of 5-40% is common for the single-stage type, with Y = 50% and N = 100 for the dual-stage type. Although these separators are very efficient, polar components and high-boilers tend to be partially adsorbed on the silicone membrane, resulting in tailing of these materials into the MS, and in some cases, complete removal of very polar materials. One definite advantage of these separators is the GC outlet is kept at atmospheric pressure with no separate pump required. A forepump is used in the dual stage mode between Stage 1 and Stage 2 to enhance removal of carrier. Routine maintenance of these silicone membrane separators is similar to other separators in that they must be kept very clean, although the extreme high-temperature/vacuum bakeout is not possible. If the yield of the separator begins to deteriorate (i.e., develops pinhole leaks), or if contaminants adsorbed on the silicone membrane begin to gradually be emitted into the MS ion source, a fresh membrane, properly conditioned by vacuum bakeout, solves these problems. It should be considered a routine maintenance procedure not to allow very polar materials to enter the separator and thus prevent adsorption on the silicone membrane. Other than the items mentioned, very little can go wrong with this type of separator provided the silicone membrane is intact.

THIN-WALLED TEFLON (LIPSKY) SEPARATOR

The molecular separator system reported by Lipsky *et al.* (10, 11), is actually a modification of the Biemann-Watson separator in that the porous glass tube is replaced by about 6-8 feet of thin-walled Teflon tubing. Both ends of the Teflon tubing are connected to stainless steel capillary tubes to reduce the pressure and allow carrier gas flow-rates of 15 ml/min or less. The Teflon tube is contained in a vacuum jacket connected to a rotary pump. The optimum operating temperature range in which the Teflon was most highly selectively permeable to helium is 250-260°C above ambient, at which yields (Y) of 80-90% are obtained, but only at an enrichment (N) of 2-6%. At temperatures 200°C above ambient there was no diffusion through the membrane; and, at 300°C above ambient, most of the helium and sample diffused through the Teflon (3). Thus, the use of this separator is somewhat limited.

Routine maintenance of such Lipsky separators consists

mainly of keeping them exceptionally clean via high-temperature (250°C) vacuum bakeouts. In addition, they should often be vacuum checked, especially where the Teflon is connected to the metal capillaries, as leaks can easily occur here and give the appearance that all materials are being passed through the Teflon in equal amounts; and, thus, a very low enrichment (N) is observed. At higher temperatures (250°C+) for extended periods of time, Teflon tends to lose its elasticity and could become loose at the metal connections. Thus, the Teflon membrane should be replaced often. This will also prevent repeated leaking of heavy materials that may be adsorbed on the Teflon into the MS ion source. Such separators are quite rugged and should require very little maintenance if adequate operating temperatures are maintained (250°C).

POROUS SILVER FRIT (BLUMER) SEPARATOR

A very small separator based on a porous silver frit to selectively remove the carrier gas has been reported by Blumer (12). This separator is mainly for use with low-flow packed columns (10-15 ml/min). The separator consists of a 1/4 inch stainless steel Swagelok tee with one arm (that is connected to a rotary pump) containing a porous silver membrane (6mm diameter, 2 mil thickness, maximum pore diameter 3 μm). The other two arms of the tee are connected to the GC and the MS ion source through a needle-type metering valve. Satisfactory analyses were obtained with saturated and unsaturated hydrocarbons (up to C_{20}) and polar aldehydes, ketones, dicarbonyl compounds and fatty acid esters (up to C_{22}). The yield (Y) of this separator is in the 30-40% range with an enrichment (N) of about 10%. The maximum temperature for operation of this separator could go as high as 400°C.

A disadvantage of this type of separator might be silver catalyzed surface reactions of organic molecules at high temperatures; however, the separator is very rugged and has an inherent low dead-volume, making it ideal for smaller diameter columns. In a recent paper, Grayson and Levy (13) report on a Blumer-type separator with dimensions optimized for use with open-tubular (capillary) columns down to 0.010-inch i.d., with flows in the 2-7 ml/min range. Yields (Y) for this microseparator range from 25-60% and enrichments (N) from 2-10% for various compounds with the separator operated at 175°C and a column flow-rate of 10 ml/min. No peak broadening or tailing was detected using this separator. A block diagram of this separator is shown in Figure 13-4.

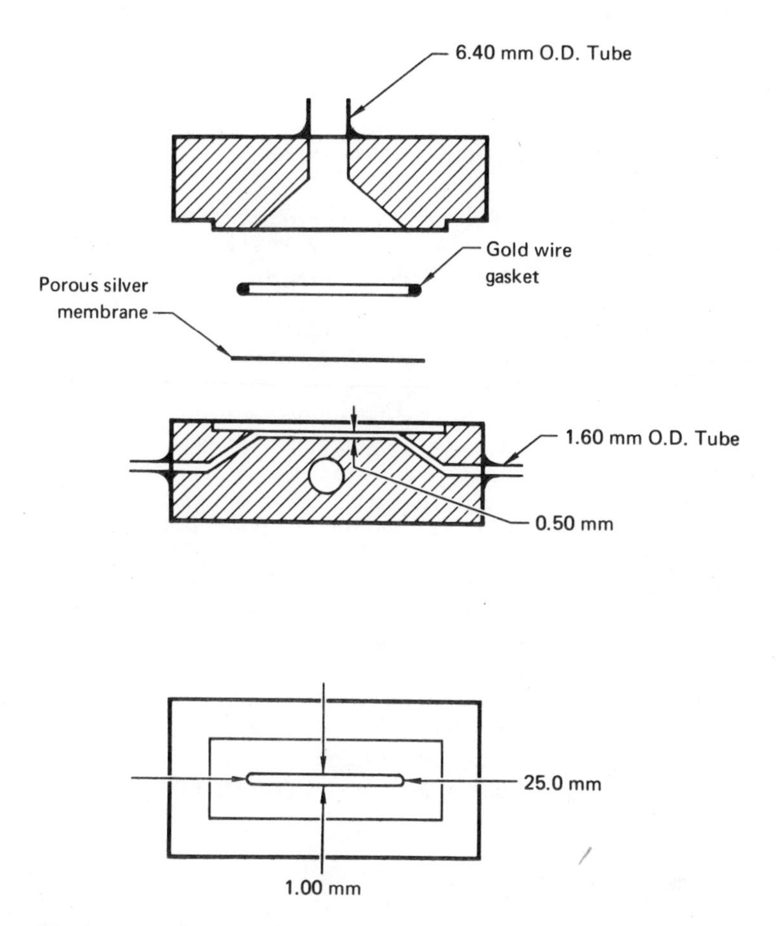

Fig. 13-4. Cross section of the microseparator (exploded view) and top view showing dimensions of separator volume. The 6.40mm O.D. tube opening is connected to a vacuum pump and the two 1.60mm O.D. tube openings are connected to the gas chromatograph outlet and mass spectrometer inlet. (Courtesy of J. Chromatogr. Sci.*) (15).*

Routine maintenance of these types of separators is quite easy due to their simplicity of construction and the fact it is very rugged. The major items are, once again, overall separator cleanliness and membrane cleanliness. Thin layers of organic materials on the silver frit can usually be removed by a high-temperature/vacuum bakeout as is used for other separators. If the GC side of the frit becomes covered with a thin coating of organic material when very low vapor pressure components are being analyzed, the frit should be replaced. Due to the simplicity of construction, problems with leaks should be minimal. In general, this type of separator can be operated for many months without any extensive maintenance required.

VARIABLE STEEL PLATE (BRUNEE) SEPARATOR

The Brunee adjustable steel plate separator allows carrier gas/sample differentiation by a variable annular slit (14). The outstanding characteristic of this separator is that it may be operated over a large range of flow rates (1-60 ml/min). The yield (Y) for this separator is 10-30% and the enrichment (N) from 10-40%, with an optimum operating temperature of about 300°C. The main disadvantage of this type of separator is possible surface reactions of organic compounds on the high temperature steel. On the other hand, the ruggedness of the separator, along with its simplicity of operation and wide dynamic range, make it a desirable unit to use in equipment where both packed and capillary columns will be employed.

Routine maintenance of this separator will generally consist of high-temperature/vacuum bakeouts. However, if the separator becomes clogged with any decomposed or high-boiling organic materials, it can be disassembled and all the stainless steel parts cleaned in a suitable solvent (acetone or chloroform), thoroughly dried and reassembled. Subsequent leak testing, and a high-temperature bakeout to remove any traces of the cleaning solvents are also in order.

MEMBRANE-FRIT (GRAYSON-WOLF) SEPARATOR

Grayson and Wolf have constructed a two-stage membrane-frit separator that is highly efficient between column flows of 5-60 ml/min (15). In addition, with this separator the GC outlet is maintained at atmospheric pressure. The yield (Y) of this separator can range from 40-60% and the enrichment (N) from 10-40%. Maximum operating temperature for this separator is 250°C due to the silicone membrane portion. Peak broadening has not been a problem with this separator.

Major maintenance items for this separator are general overall cleanliness and checks on the condition of the silicone membrane. If the membrane develops pin-hole leaks (as sometimes will happen after extended operation at high temperatures), the first stage will not be operating as a separator at all, and the total separator performance will deteriorate. If the efficiency of this separator begins to decrease, the first check point is the silicone membrane.

CONCLUSIONS CONCERNING ROUTINE MAINTENANCE AND OPERATION OF MOST MOLECULAR SEPARATORS

As shown by Grayson and Wolf (4) and discussed in the previous section, several GC factors can affect separator operation and, as a result, affect MS resolution. The most prominent are: 1. Carrier flow-rate--the separator must be operated in the correct range. 2. Carrier/sample ratio--if the separator is not being operated in its area of greatest efficiency, a high ratio will exist which generally has a detrimental effect on MS resolution and sensitivity. 3. Peak elution width--if a peak is too sharp, the rapidly changing sample pressure will cause mass spectral resolution anomalies. On the other hand, if the peak is too wide, it is almost impossible to obtain a representative spectrum due to lower than optimum sample/carrier gas ratio. 4. Impure peaks containing more than one component--they may give excellent MS resolution but a confusing spectrum. The experienced operator can usually spot a mixture from its mass spectrum. However, probably the most important GC condition to be maintained for proper separator performance is to keep the column flow-rate (Item 1) well within the limits prescribed for any given separator.

Generally, for a good separator performance, the unit should be operated within the specification given for the individual separators in the previous section of this chapter; otherwise, poor overall performance will result. It should be stressed that the separator and all connecting tubing and valves should be heated to the maximum extent the components in the various samples will allow, and these temperatures should be checked periodically to detect any possible cold spots.

In our experience, capillary columns have been operated with the result that 50-90% of the effluent goes to the molecular separator/MS system while the remaining 10-50% is directed to a GC detector (usually an FID). Thus, we obtain the standard chromatogram along with mass spectra of desired peaks. Some newer units rely on the total ionizing current in the MS for the chromatogram, and the total column effluent is directed to the separator. However, one must be careful when directing the total column flow to the separator not to exceed the optimum flow-rate range of the separator and install a split valve if necessary (especially with packed columns). An obvious malfunction that can be devastating to the performance of a separator system is a leak somewhere between the column exit and the MS ionization source. If a leak is present, air and other contaminants will be sucked in and pass into the MS ionization region resulting in high background level. Also, to keep the background level down after a leak has been tracked down and terminated, the separator system should be baked out at a high temperature under vacuum as described previously. Several good high-temperature vacuum leak sealants are commercially available, with the aerosol variety being particularly effective in stopping leaks found in the tube fittings used with most GC-separator-MS installations. The best procedure is simply to spray a given joint with sealant, and observe the mass spectrum for declining intensity of the water series (m/e 16-18), nitrogen (m/e = 28), oxygen (m/e 32), argon (m/e = 40), and CO_2 (m/e = 44).

These are only a few suggestions to keep the GC/MS system operating at top efficiency. The items to remember to keep a separator system operating most effectively are:

1. Do not exceed the specified flow-rates for the separator.
2. Keep it as warm as possible.
3. Keep it clean by periodic vacuum bakeouts (see following paragraph).
4. Be sure to use the correct system for the type of compounds being analyzed to prevent loss or sample decomposition (thermal or metal catalyzed).
5. Make sure the system is leak-tight, as "air pollution" of the GC effluent can cause high MS background.

If these general rules are followed, the mass spectrometer should yield invaluable compositional data for separated components on a routine basis.

Sometimes GC/MS interfaces become coated with silicone from analyzing silylated derivatives and from column bleed from silicone-type packings. These deposits are difficult to remove, and how they are removed depends upon the type of molecular separator employed. If a glass or stainless steel separator is used, it may *sometimes* be cleaned in situ by switching the carrier gas to hydrogen and raising the temperature of the separator to at least 350°C, then, making repeated injections of standard methanolic - BF_3 solution that is used to clean silicone deposits from flame ionization detectors. The presence of pure hydrogen, high-temperature, silicone and BF_3 will probably form HF on the reactive surfaces (frits) usually present in separators of this type to form volatile silicone tetraflouride. However, due to the absence of the heat of the hydrogen flame, it will take many more injections of the methanolic-BF_3 solution to free the separator of silicone deposits. If this does not clean the silicone deposits from the detector, they usually can be eliminated by removal of the separator from the GC/MS system and cleaning it in hydrofluoric acid, hopefully forming SiF_4 during the cleaning. This cleaning can be aided by first heating the separator to about 350°C and passing pure O_2 through it to hopefully convert most of the silicone present to SiO_2, which will make it more reactive with respect to the HF via the following reaction:

$$SiO_2 + 4HF \rightarrow SiF_4 + 2H_2O$$

AUTOMATIC SAMPLING SYSTEMS

Most of today's automated and computerized chromatographs can easily be fitted with an automatic liquid sampling system. Automatic liquid samplers are available from essentially all of the major GC-unit manufacturers, and all samplers work in essentially the same way, i.e., pneumatic syringe drive and positioner, with electronic logic supplied either by a controller console or, if a minicomputer is used, by the computer software. Samples are usually placed in crimped-top vials fitted with Teflon-backed silicone-rubber septa. These vials fit into a carousel-type assembly that can hold up to 36 separate vials. The sampler operation logic usually allows several syringe flushes to clean out traces of the previously-run sample. Then, the syringe is pumped a few times to rid the sample of air bubbles, the syringe needle withdrawn from the vial, the syringe flipped pneumatically to the injection position, the sample injected automatically and the needle withdrawn. The sampler then remains inactive until the logic controller receives a signal (either manual or automatic) to start another run. At this time, the next vial in the carousel is moved to the inject position, and the syringe needle pneumatically inserted into the vial through the silicone rubber septum, after which the sequence of events described above (flush, sample and inject) are repeated for this sample. With most basic automatic sampling systems, the operator may choose the number of full-syringe flushes for the series of analyses and the sample size (up to about 10μl), as well as the number of samples to be analyzed. With the new computerized logic programs available, the vials in the sample carousel do not have to be analyzed in series, but the unit can be programmed to index any vial into the sampling position at the start of an analysis.

Since automatic liquid samplers are a combination of both mechanical and solid-state electrical devices, there are only two types of things that can go wrong. The sampler can receive the wrong electrical signal from the controller, causing operation of the logical sequance of events to go awry. On the other hand, all of the mechanical parts are usually pneumatically powered, actuated by the controller logic. Thus, if a given part of the cycle malfunctions, it is either caused by no signal, or the wrong signal, from the electronic controller (for example, a microswitch may not be closing, as programmed); or, if the correct electrical impulse is present and a malfunction occurs, it is easily traceable to some malfunction of the pneumatic-mechanical system. We have had direct experience with a widely used automatic sampler for about 2-1/2 years and have only had one problem, and that was the shorting-out of a capacitor on one of the logic-control circuit boards in the controller unit.

For running routine analyses, such as true-boiling-point GC simulated distillations, such automatic samplers are invaluable in the amount of operator time they can save, as they can be routinely set up to run uninterrupted on a 24-hour basis by simply adding samples to the sample carousel whenever necessary.

Thus, in this chapter we have discussed some of the more common auxillary techniques used for qualitatively monitoring GC column effluents (other than the conventional GC detectors), and have endeavored to point out some routine maintenance procedures that should aid in the optimum operation of these auxillary systems.

REFERENCES

1. J. T. Walsh and C. Merritt, *Anal. Chem., 32,* 1378 (1960).
2. R. S. Gohlke, *Anal. Chem., 31,* 535 (1959).
3. D. E. Rees, *Talanta, 16,* 903 (February 1969).
4. M. A. Grayson and C. J. Wolf, *Anal. Chem., 39,* 1438 (1967)
5. J. T. Watson and K. Biemann, *Anal. Chem., 36,* 1135 (1964).
6. *Ibid., 37,* 844 (1965).
7. R. Ryhage, *Anal. Chem., 36,* 759 (1964).
8. R. Ryhage, *Arkiv. Kemi., 26,* 305 (1967).
9. P. Llewellyn and D. Littlejohn, *Varian Technical Quarterly,* p. 6 (Spring 1966).
10. S. R. Lipsky, C. G. Horvath, and W. J. McMurray, *Anal. Chem., 38,* 1585 (1966).

11. S. R. Lipsky, C. G. Horvath, and W. J. McMurray, *Gas Chromatog,* 1966, A. B. Littlewood, p. 299.

12. M. Blumer, *Anal. Chem., 40,* 1590 (1968).

13. M. A. Grayson and R. L. Levy, *J. Chrom. Sci., 9,* 689 (1971).

14. C. Brunee, *et al.,* 17th Annual Conference on Mass Spectrometry and Allied Topics, Dallas, Texas (May 1969), Paper No. 46.

15. M. A. Grayson and C. J. Wolf, *Anal. Chem., 42,* 426 (1970).

Chapter 14

GAS CHROMATOGRAPHIC COMPREHENSIVE TROUBLESHOOTING

INTRODUCTION

The most expensive GC instrument is worthless if it does not operate properly. Most intruments are so complex and so many variables are involved that it may not be easy to determine the cause of the difficulty. Short of simple luck, the best method of attacking the problem is a logical one in which each possible cause is eliminated in the order in which they are most likely to occur. In the following pages GC troubleshooting is brought into a simple and quick form.

The tables are divided into several categories descriptive of the symptoms found frequently in ailing chromatographs. Each category is followed by three sections designed to pinpoint the problem and find a remedy. The first section defines the symptom and, where pertinent, provides sample chromatograms to illustrate the problem. Following this section is a section entitled "Accompanying Symptoms." Often more than one symptom is present providing a clue as to the origin of the problem which is causing these symptoms. In this table, the entire GC system is broken down into four dividions as follows:

1. Gas System (Gas): The gas system, here, will include all GC parts involved in moving the sample through the column. The tubing, gas supply, flow-meters, etc. are included.

2. Column (Col): The column includes the column tubing, its packing and/or liquid phase, as well as the injector or sampling valves.
3. Detector (Det): The detector is considered exclusive of the electrometer, bridge, and power supply. Thermal Conductivity Detectors (TCD), Flame Ionization Detectors (FID), and Electron Capture Detectors (ECD) are considered in particular. Included with the detector are GC parts involved in transporting detector gases (H_2, air) to the detector.
4. Electronics (Elec): Electrometers, power supplies, bridges, recorders, and oven and detector heaters are included here.

If an accompanying symptom is present, the most likely system to troubleshoot and the order in which to troubleshoot the other systems are given.

In the third section a table is given which lists the probable causes of the problem indicated by the symptoms. Where, as in drift, the chromatogram may take on more than one shape, Column 1 indicates the type of chromatogram evidenced. Column 2 lists the system where the problem originates. Causes of the symptom are listed in the next column followed by methods of isolating and confirming the source of the difficulty. Means by which the problem can be remedied are given in the next column, though the remedy is often obvious once the problem is isolated. In the final column, a reference page or pages is given to direct the reader to a more detailed analysis of the subject.

In the trouble-shooting tables, the systems are dealt with in the order in which they occur in the GC instrument. The most likely cause of the problem is dependent upon many variables such as the type of sample being analyzed, the type of column being used, and, perhaps most important, the individual "quirks" of the instrument.

Baseline Problems: *Drift*

A. Rising Baseline	Falling Baseline (Isothermal Operation)
B. Rising Baseline	(Temperature Programming) Falling Baseline
C. Irregular Drift	(Isothermal Operation)

Definition

Drift is characterized by changes in the baseline. Drift is generally made up of slow fluctuations in the baseline when no sample has been injected. The baseline may either rise or fall steadily as in A and B (page 280) or it may fluctuate as in C.

Drift: Accompanying Symptoms

	Order in which to Troubleshoot System			
Accompanying Symptoms	Gas Flow	Column	Detector	Electronic
Poor retention time reproducibility	1	2	--	--
Noise	2	1	3	4
Noise with sensitivity loss	1	--	2	3
Drift increases or decreases with column temperature	2	1	3	--
Drift decreases with time	2	1	3	--
Detector temperature fluctuation (TCD)	--	--	1	--
Drift at attenuation	--	--	--	1

Baseline: Drift Problems

Drift Type	System	Cause	Isolate and Confirm	Remedy
A or C	Col	Column bleed	Drift decreases with temperature decrease.	Recondition column at near the maximum liquid phase temperature for one hour or until drift disappears (disconnect column from detector), Chapter 10. Use dual column system (see Chapter 10, columns).
	Col	Contamination of column		Recondition and/or replace column, Chapter 10.
B	Col	Unstable column oven	Monitor oven temperature.	Check for proper oven fan operation and air flow, Chapter 10.
				Replace temperature controller, Chapter 10.
	Elec	Defective electrometer.	Disconnect detector cables.	If drift continues, replace electrometer, Chapter 12.
			Check electrometer batteries.	Replace batteries every three months. Chapter 12.
B	Elec	Unstable TCD detector temperature.	Monitor detector temperature.	Replace temperature controller, Chapter 11.
C	Gas	Poor H_2 gas regulation (FID)	Check H_2 flow at detector for proper and constant flow.	Readjust flow or replace flow controller, Chapter 8. Leak check, Chapter 8.
C	Gas	Carrier gas leak	Leak check all fittings.	Tighten or replace fittings.
		Poor carrier gas regulation	Monitor gas flow at column outlet.	Be sure supply tank is maintained at a sufficient pressure to ensure proper operation of flow controllers.
				Repair or replace flow controllers.

Baseline: Drift Problems - Cont'd.

Drift Type	System	Cause	Isolate and Confirm	Remedy
C	Det	Poor instrument location	Stop room air condition system and check drift.	Move instrument away from heating and cooling outlets or other drafts which might affect temperature of the instrument environment.
C	Det	Detector contamination	Reduction in detector temperature should decrease drift.	Clean detector according to Appendix A, Detector Cleaning.
C	Elec	Instrument not grounded	May be accompanied by noise.	Ensure instrument and recorder securely attached to a good earth ground.
	Elec	Electrometer defective (FID, ECD)	Isolate electrometer from detector to see if drift continues.	Check zero circuitry, batteries balance circuit.
	Elec	Electrometer not warmed up	Drift will discontinue when instrument has had time to warm up.	Warm up time may require as long as 24 hours.
	Elec	Bad bridge (TCD)	Movement of bridge wires increases or decreases drift.	Check power supply, zero pots, etc.
D	Gas	Carrier gas supply pressure too low	Monitor flow at outlet of column.	Increase carrier gas supply to pressure regulator and flow controllers.
	Col	Unsteady column oven temperature	Check oven seals.	Make sure oven is sealed properly and proper oven insulation is in place.

Baseline: Noise

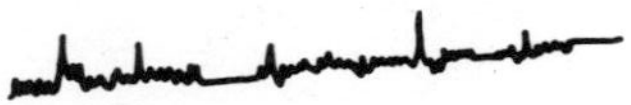

Noise Type	System	Cause	Isolate and Confirm	Remedy
A	Gas	Dirty injector	Low injector temperature should decrease noise.	Clean injector and replace septum.
		Leak	Leak Check at all connections.	Tighten or replace connections.
		Carrier gas flow too high	Check flow at column outlet.	Readjust flow controllers to obtain optimum flow rate
		Contaminated carrier gas	Replace carrier gas supply.	Discard contaminated carrier gas.
A	Col	Contaminated column	Noise should decrease at lower temperatures.	Recondition column for at least one hour at slightly less than the maximum temperature of the liquid phase. It may be necessary to replace a badly contaminated column.
		Column bleed	Noise should decrease at lower temperatures.	
A	Det	Faulty detector cables	Disconnect cable at electrometer. If detector has checked out O.K. and noise discontinues, cable is at fault.	Replace cables.
		Contaminated air or hydrogen (FID)	Replace hydrogen and air supply.	Discard contaminated gas supplies.
	Det	Improper hydrogen flow (FID)	Check hydrogen flow at flame tip.	Adjust hydrogen flow as indicated in the instrument manual.
		Improper air flow (FID)	Check for proper air flow rate.	Adjust air flow to proper level.
		Water condensing inside detector (FID)	Check detector temperature.	Raise detector temperature to slightly higher than 100°C.

Noise Type	System	Cause	Isolate and Confirm	Remedy
A	Det	Contaminated detector	Lower detector temperature (but not below 100°C). Noise should decrease.	Clean detector as outlined in Dectectors Chapter.
		Dirty detector insulators (ionization detectors)	Noise should decrease at lower temperatures.	Clean with a residue-free solvent. Do not handle clean insulators.
A or B	Elec	Loose cable connections	Check all plugs and screw connections to make sure they are solid.	If necessary, clean or replace bad connectors.
	Elec	Malfunctioning integrator feedback	Bypass integrator in system.	Change cable connections or consult instrument manual.
	Col	Column packing entering detector	Change column.	Increase flow to remove loose particles. Apply back pressure through detector (disconnect column).
A or C	Elec	Bad ground	Check to see that ground connection is secure.	Be sure that the ground connection is attached to a good earth ground.
		Dirty switches	Examine switches for heavy oxidation and firm seating.	Clean switch contacts with either a light grit emery paper or contact cleaner.
		Dirty recorder slide-wire	Noise will probably be intermittent, occurring always at the same point on the recorder scale. Noise will not change with attenuation.	Clean slide-wire according to recorder manual. Special slide-wire cleaner is available for this purpose.
		Recorder not operating properly	Short recorder input. If noise continues, recorder is source.	Check recorder damping and gain control. Consult recorder manual for further troubleshooting.
C	Gas	Hydrogen generator H_2O overflow	Check hydrogen generator for H_2O overflow.	Dry transfer lines from generator to column with heat or replace.

Noise Type	System	Cause	Isolate and Confirm	Remedy
C	Elec	AC input circuit	Remove other loads from circuit into which GC is hooked.	Reduce circuit load or change AC input circuit for GC.

* * * * * *

Types of Noise: A. Irregular or Intermittent Noise
B. Regular Noise
C. Spikes

Definition: *Noise* is characterized by a jerky recorder pen. At times, as in Type A (above), the noise is erratic and may disappear completely for short periods of time. A second type of noise, shown in B, is a regular pattern of spikes. Noise is not necessarily caused by the recorder or any other segment of the electrical system, and often it is caused by contamination of the column or gas system.

Accompanying Symptoms: These additional symptoms may provide a guide or direction to the origin of the problem. The system in which the problem originates is listed in the troubleshooting tables which follow. The most logical order of troubleshooting is indicated.

Accompanying Symptom	Gas Flow	Column	Detector	Electronic	Remarks
Drift (less at lower temperature)	2	1	2	--	Most likely column bleed.
Noise and drift decrease with time	3	1	2	--	Contamination.
Sensitivity loss	--	--	1	2	Wrong flow with ionization detectors.
Poor retention time reproducibility	1	2	--	--	Common with temperature programming.
Drift at attenuation (short)	--	--	--	1	Check out recorder.

Baseline Problems: Zeroing

This problem arises when the recorder cannot be set at zero. The cause of zeroing difficulties are varied; and, while it is not a frequent problem, zeroing difficulties may be the most difficult to remedy.

Baseline: Zeroing

System	Cause	Isolate and Confirm	Remedy
Gas	Improper flows (carrier, air, H_2).	Check flows with bubble meter.	Reset flows according to instrument manual.
Col	Excessive column bleed.	Lower column temperature, zero should go down.	Recondition column at near maximum liquid phase temperature until bleed stops.
Det	Dirty detector (particularly FID and ECD).	Disconnect cables at detector and attempt to zero.	Clean as outlined in Detectors Chapter.
	Weak filament (TCD).	Disconnect cables at detector and attempt to zero.	Replace filaments.
Elec	Recorder zero not set.	Short recorder input and check zero.	With input still shorted, set recorder zero.
Elec	Malfunctioning electrometer.		Consult electrometer manual.
	Improper recorder connection.	Check all recorder connections for proper positioning.	Connect cables to recorder as outlined in instrument manual.
	Defective recorder.	Short recorder input and check zero.	If zeroing not possible, consult instrument manual for troubleshooting.
	Bad battery.	Check with voltage meter.	Replace batteries.

Baseline Problems: Cycling

Cycling is a baseline which shows a regular wave pattern. The problem is most likely caused by fluctuations in the carrier gas flow rate. In the case of TCD detectors, the problem can be caused by variations in the detector cell temperature.

Baseline Problems: Cycling

System	Cause	Isolate and Confirm	Remedy
Gas	Carrier flow variation.	Monitor flow at detector.	Flow controllers require high inlet pressures (30 psi or above). Replace faulty flow controllers.
	Pumped gas supply.	At high sensitivities, variations in supply pressure from pumped gas supplies will cause cycling.	Use a ballast tank to eliminate pressure variations.
Det	Cycling detector temperature.	Monitor detector temperature. Check line voltage supplying instrument.	Repair or replace faulty detector temperature controller.
	Condensation in TCD detector.		Increase flow rate through detector. If necessary, clean detector as outlined in Chapter on Detectors.

Baseline Problems: Spikes

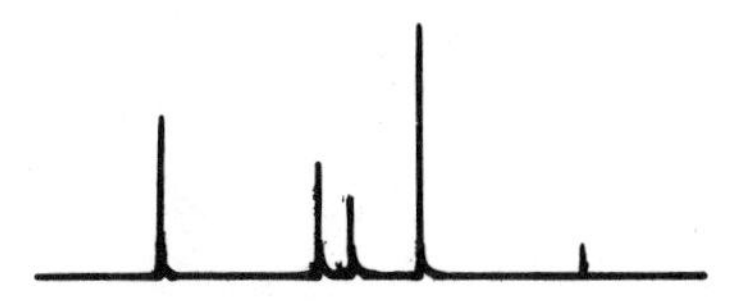

Unlike noise, spikes often travel nearly completely across the recorder range. Spikes are generally intermittent and vary in size. Usually caused by dirty electrical switches and connections or a contaminated detector, spiking may also result from line voltage fluctuations or drafts.

Baseline: Spikes

System	Cause	Isolate and Confirm	Remedy
Gas	Contaminated gas.	Connect gas supply directly to detector.	Replace gas supply.
Col	Loose column packing.	Tap column sharply and observe baseline.	Disconnect from detector and increase flow rate.
Det	Drafts.	Check for drafts from air conditioners, blowers, opening doors.	Move instrument.
Det	Dirty venting tube (ECD)		Clean as outlined in Chapter 10 on detectors.
Elec	Dirty electrical connections (FID, ECD).	Spikes will continue with electrometer in "Balance" position.	Clean connections, switches, pots and troubleshoot power supply.
Elec	Zeroing circuit.	Spikes will stop with electrometer in "Balance" position.	Consult instrument manual for troubleshooting zero circuitry.
Elec	Dust, dirt in detector (FID).	See remedy.	Blow a clean, dry air supply across detector.
Elec	Line voltage fluctuations.	Monitor line voltage coming into instrument.	
Elec	Vibrations.	Usually aggravates poor electrical connections.	Remove source of vibrations (i.e., oven fan) or isolate instrument from shock.

Distorted Peaks: Ghost Peaks

Types of Ghost Peaks

A. Ghost Peak

B. Ghost Peaks

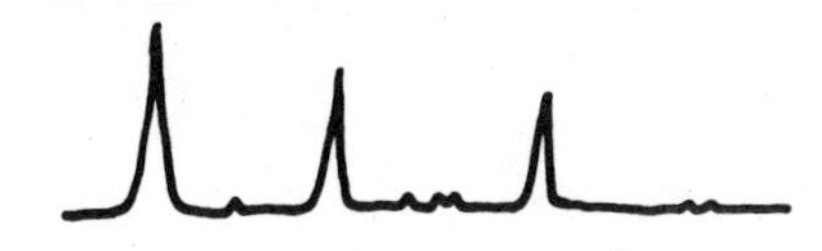

C. Ghost Peaks

Often during the course of an analysis, peaks will appear which do not fit the expected pattern of the chromatogram. These unexpected peaks, known as Ghost Peaks, are many times not well resolved (A), are generally present in small quantities (B), and sometimes resemble noise as they emerge together with the larger peaks of the chromatogram (C). The cause of these ghost peaks is most often found in the inlet system, the gas system, or the column system.

Ghost Peaks

Type	System	Cause	Isolate and Confirm	Remedy
B, C.	Gas	Contaminated sample.	Obtain another sample. Be sure to clean syringe thoroughly.	Use purest available samples.
A, B, C	Gas	Leak.	Leak check all fittings at maximum temperature at which they are used.	Tighten or replace work fittings.
A, B, C	Col Gas	Contamination.	Lower system temperatures and observe effects.	Clean or replace contaminated system components.
		Septum bleed.	Note maximum recommended temperature for septum. Cover new septum with aluminum foil.	Pre-condition septums at operating temperature. Replace worn septums.
A, B, C	Col	Reaction of sample with column liquid phase.	Inject a different type of sample.	Use a different column.
		Reaction of sample with other GC parts (Septum, column tubing, fittings, etc.)	If sample composition is known, examine reactivity with GC system components.	Where possible, use less reactive parts made of quartz glass, Teflon, stainless steel.
A	Col	Heavy M.W. material from previous analysis is eluting.		Condition column at near maximum temperature for about one hour or until a steady baseline is observed.
	Col	Condensed materials (H_2O or other impurities) eluting during temperature programming.		Install or replace carrier gas filter.

Ghost Peaks - Cont'd.

Type	System	Cause	Isolate and Confirm	Remedy
B	Gas	Air peak (not present with FID).	Air peak is usually the first peak to emerge.	Cannot be eliminated with making syringe injections, if necessary, air peak can be eliminated by using a sampling valve.
B	Col	Column liquid phase desorption when injecting a solvent.	Inject a sample which does not include solvent.	Inject solvent into column repeatedly and recondition column.
C	Col	Decomposition of sample.	Lower injector temperature.	Lower injector temperature (if high enough to cause thermal decomposition). Use different column liquid phase (if phase is not compatible with sample).

Distorted Peaks: Others

Types of Distorted Peaks

A. Leading Peak

B. Tailing Peak

C. Flat Top

D. Round Top

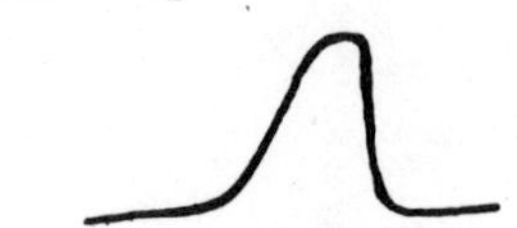

E. Negative Dip before Peak

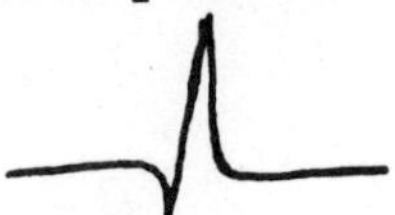

F. Negative Dip after Peak (ECD)

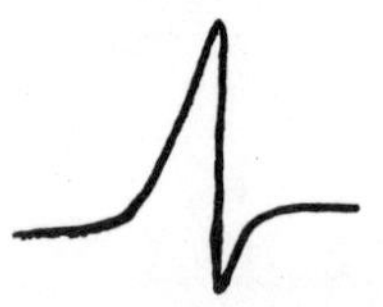

The ideal chromatogram contains all very nearly Gaussian shaped peaks. Other peak shapes encountered are usually caused by improper operating conditions. Column overload is the most common cause of almost all types of distorted peaks.

Distorted Peaks

Type	System	Cause	Isolate and Confirm	Remedy
A	Col	Column overload.	Inject smaller sample.	Use smaller sample. Use larger column.
		Sample condensation.	Check vapor temperature of sample.	Use high enough injector and column temperature.
		Two peaks eluting simultaneously.	Lower column temperature about 30°C to try for better separation.	Optimize column operation conditions. Use a more suitable column.
	Gas	Poor sample injection.	Inject another sample.	Use proper injection techniques as outlined in Appendix B.
		Flow through detector too slow.	Check flow at detector outlet. Carrier flow of at least 30 ml/min should be entering detector.	Use scavenger gases at end of column where necessary to increase flow of sample into detector.
	Col	Injector temperature too low.	Raise injector temperature.	Use high enough injector temperature to readily volatilize the sample.
		Interaction with column support material.	See remedy.	Use a less active support. Increase liquid loading of column. Derivatize sample.
		Column overload (with gas samples).	Inject smaller sample.	Use smaller sample. Use larger column.
		Dirty injector tube.	Check for sample residue. Increased injector temperature.	Clean injector tube by flushing with solvents.

Distorted Peaks - Cont'd.

Type	System	Cause	Isolate and Confirm	Remedy
B	Col	Two compounds eluting together.	Lower column temperature to improve separation.	Optimize column operating conditions (flow, temperature). Use different column (liquid phase diameter, length).
		Oven temperature too low.	Increase oven temperature.	Use higher oven temperature if resolution remains good.
C	Elec	Electrometer saturated.	Switch electrometer.	Use smaller sample. Use different electrometer range.
		Recorder range exceeded.	Switch recorder range.	Use smaller sample. Use different recorder range.
D	Det	Operating outside of linear range of detector (most likely with ECD).	Examine linear range of the detector in use.	Reduce sample size.
	Elec	Recorder gain set too low.	See remedy.	Adjust recorder gain to proper setting.
	Gas	Pressure surge.	Often seen before large peaks.	Not easily eliminated. Try: 1. Reduce sample size; 2. Reduce column bleed. 3. Check for contaminated detector.
F	Det	ECD sample overload.	See remedy.	Allow time for baseline to return to normal. Dilute sample or reduce size.
		ECD, dirty detector.	See remedy.	Clean detector as outlined in the Chapter on detectors.

Reproducibility Problems: Retention Time

Most uses of chromatography require that the retention times of peaks be reproducible. Non-reproducibility may develop suddenly or come about as a result of gradual changes. In isothermal GC, the most common cause of retention time difficulty lies within the flow system, followed closely by the column system. In temperature programmed GC, by far the most common cause of differing retention times is non-reproducible temperature profiles.

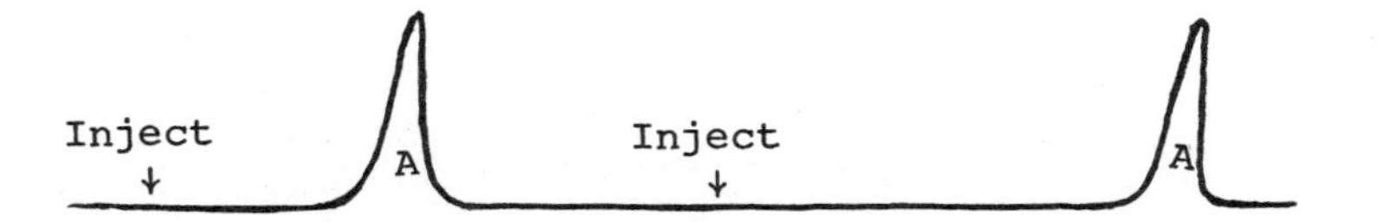

Accompanying Symptoms: These additional symptoms may provide a guide to the origin of the problem.

Accompanying Symptom	System				Remarks
	Gas Flow	Column	Detector	Electronic	
Noise	1	2	--	--	Probably a leak in carrier gas.
Sensitivity	1	2	--	--	Probably a leak in carrier gas.
Drift	1	3	--	--	
Distorted Peak	1	2	--	--	Sample size too large is another possibility
Gradual Decrease in Retention Time	2	1	--	--	

Reproducibility: Retention Time

System	Cause	Isolate and Confirm	Remedy
Gas	Insufficient retention.	Peak measured should require at least four times as long to elute as does the "air" peak.	Lower temperature of column.
			Lower flow rate through the column.
	Poor injection.	Change method of syringe handling and/or syringes.	Syringe injection should be accomplished quickly and cleanly as detailed in Appendix B.
	Leak	Replace septa	Septums should be replaced periodically; especially during high temperature operation or frequent injections.
			Leak check all fittings.
	Changing flow (while temperature programming).	Check flow at column outlet at both minimum and maximum temperature.	Flow rates should not differ by more than 2 cm^3/min for a 1/8 in. I.D. and larger columns.
	Bad flow control.	Monitor flow rate at column outlet.	Increase flow controller inlet pressure.
Col	Column temperature not equilibrated.		Allow five minutes for temperature to equilibrate after reaching operating temperature.
			When temperature programming, allow time to equilibrate at starting temperature. More time is required when starting near ambient.

Reproducibility: Retention Time - Cont'd.

System	Cause	Isolate and Confirm	Remedy
Col	Bad column temperature control.	Check oven seal.	Make sure oven is sealed properly before beginning.
		Monitor column temperature.	Troubleshoot temperature programmer (gears, and thermal couple).
	Too much sample.	Peak tailing will result.	Dilute sample or reduce amount injected.
			Go to larger column.
	Operation outside of range of column liquid.	Consult Appendix I for proper range.	Generally, do not operate within 5°C of either minimum or maximum temperature.
	Column deterioration.	Probably accompanied by loss of resolution.	Replace column.

Reproducibility Problems: Sensitivity Changes

Within the limits of injection techniques, the relative peak areas should be the same in successive analysis. Most changes in sensitivity are actually changes in the amount of sample introduced into the column brought about by improper injection techniques. Even with injection techniques optimized, absolute peak areas will vary by around 5% at best. Another frequent cause of sensitivity change (loss) is a leak in the gas system. Changes in carrier gas flow will be inversely proportional to the sensitivity of ECD and TCD dectecors.

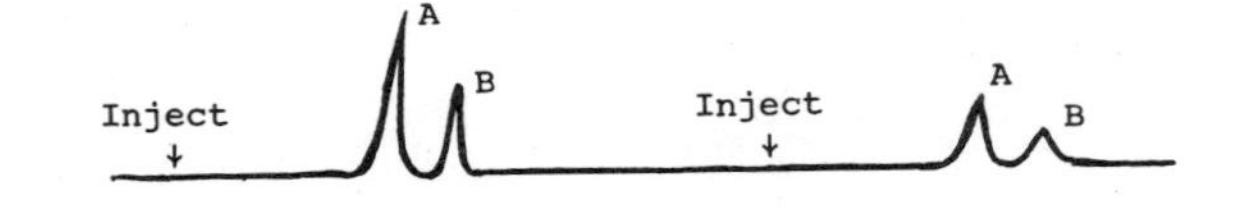

Accompanying Symptoms

Accompanying Symptom	Suspected System				Remarks
	Gas Flow	Column	Detector	Electronic	
Drift	1	2	3	--	Leak check.
Noise	1	2	--	--	Leak check.
Baseline not returning to zero	--	--	2	1	Check recorder zero.
Ghost Peaks	2	--	1	--	Bleed or contamination
Round Top Peaks	--	1	--	--	Overload detector.
Changes in retention time	1	--	3	2	Changes in carrier gas flow with ECD, TCD.

Sensitivity Changes

System	Cause	Isolate and Confirm	Remedy
Gas	Injection technique.	Repeat analysis. Frequent rising and falling of peak area indicates poor injection techniques.	Proficiency in the use of a syringe requires practice. See Chapter for proper methods of injection. Maximum repeatability is about 5% with the best syringes.
	Carrier gas flow changes (ECD, TCD).	Check carrier gas flows.	Readjust flow or replace flow controller if needed.
	Changes in H_2 flow (FID).	Check H_2 flow at flame tip.	Optimize H_2 flow (normally in approximately 1 to 8 ratio H_2 to air).
Gas	Leak.	Measure all flows at the detector and leak check if necessary.	Tighten or replace connections.
Col	Faulty syringe.	Use a different syringe if available.	Replace syringe needle, plunger, seals, etc. as needed or return to manufacturer for repairs.
	Inconsistent sample preparation.	Review sample preparation procedures for possible inconsistencies.	Use as few steps as possible in preparing sample. Hold all conditions constant. Take extreme care to avoid sample contamination from syringes, sample vials, atmosphere, etc.
Det	Improper collector alignment.	Check alignment of collector with respect to flame jet and igniter.	Consult instrument manual for proper alignment.

Sensitivity Changes - Cont'd.

System	Cause	Isolate and Confirm	Remedy
Det	Dirty detector (FID).	Probably accompanied by noise.	Clean as outlined in Chapter 11.
	Dirty filaments (TCD).	Accompanied by increased zero.	Clean detector as outlined in Chapter 11.
	Short in cell (ECD).	Check anode lead to ground voltage. It should read 2-4 volts. If less, disconnect cable and recheck at the connector. If now normal, short is indicated. If voltage still low, troubleshoot pulser.	Replace cable and examine detector for possible shorts.
	Detector overload.	Rounded top peaks will probably be evidenced.	Stay with linearity of detector. Reduce sample size.
Elec	Varying column temperature.	Will be accompanied by changes in retention time.	Repair or replace temperature controller or programmer.
	Recorder settings changed.	Check recorder damping and gain.	Readjust recorder damping and gain.
Elec	Low collector electrode voltage (FID)	Extinguish flame, measure collector voltage.	See instrument manual for appropriate collector electrode voltage.
		Check battery voltage.	Replace batteries.
	Faulty pulser (ECD).	A. Zero recorder pen at 5 μsec pulse interval and sensitivity of 100 x 8 (tritium) or 100 x 8 (nickel). B. Slowly increase pulse interval to 150 μsec and return. C. Pen should reach midway across recorder and return. D. If not, pulser is faulty.	Consult instrument manual for repair procedure.

Other Problems

Not all problems are connected to the chromatogram. Some difficulties are evident just by checking currents and temperatures of various systems within the instrument. These problems can occur in heaters, detector filaments, attenuators, etc. There is little difficulty with isolation and confirmation with these problems, and the remedies are generally obvious. For these reasons, we shall depart from the previous format and simply list the problems together with their probable causes.

A. Oven will not heat (detector or column).

1. Defective switch
2. Loose wiring
3. Defective powerstat
4. Blown fuse
5. Defective upper limit switch
6. Open heater element
7. Open programmer interconnect

B. Injector or collector will not heat.

1. Heater element open
2. Powerstat or controller defective
3. Broken wire

C. Heater on but pyrometer will not indicate temperature.

1. Defective thermocouple wire or connection
2. Defective selector switch
3. Defective pyrometer (if all selections show no temperature)

D. Flame goes out (FID).

1. Restriction in flame tip
2. Loss of hydrogen supply
3. Loss of air supply
4. Carrier gas flow too high
5. Sample too large

E. Ignitor will not ignite burner (FID).

1. Broken ignitor coil
2. Defective ignitor voltage cable
3. No voltage out of the electrometer

F. No filament current (TCD only).

1. No output from power supply
2. Defective switch
3. Blown fuse in power supply section
4. Open current meter
5. Open wiring
6. Filaments burned out

G. Will not attenuate properly (all detectors).

1. If faulty at only one step,
    a. Check resistor for that step.
2. If faulty at all steps,
    a. Check recorder electrical zero.
    b. Check resistor common to all steps.
    c. Check recorder hookup.

Other Problems: No Peaks

One of the most disappointing occurrences in GC is injecting a carefully prepared sample and waiting patiently for the peaks to emerge only to have none appear. The most common causes of no peaks are the detector power being off, the flame not being ignited (FID), the carrier gas leaking, and a bad syringe. This problem is also related to "Reproducibility Problems: Sensitivity Changes" so consult that section also.

No Peaks

System	Cause	Isolate and Confirm	Remedy
Gas	No carrier gas flow.	Check tank pressure and column outlet flow.	Turn on carrier gas flow and set to proper flow rate.
	Leak.	Leak check all fittings. Some small peaks may show up.	Tighten or replace defective fittings.
	Leaking septum.	Replace septum.	Change septums frequently depending on the injector temperature.
	Hypodermic syringe defective.	Check syringe delivery into a piece of dry tissue paper.	Repair or replace syringe.
Det	Cold injector flame not ignited (FID).	Check injector temperature. Check recorder zero or check for H_2O condensate above flame.	Increase temperature of injector. Re-ignite flame. Cleaning of ignitor may be necessary. Also check H_2 flows at flame tip.
	No detector power.	Check switches.	Turn detector switches to on position.
	No cell voltage being applied to detector (ionization detectors).	Monitor cell voltage if possible. Check switches.	Turn switch on. Troubleshoot power supply.

No Peaks - Cont'd.

System	Cause	Isolate and Confirm	Remedy
Elec	Recorder improperly connected.	Check recorder cables for proper orientation.	Hook recorder cables to proper position.
	Recorder or electrometer attenuated too high.	Decrease attenuation and check for peaks.	Operate recorder and/or electrometer at a more sensitive position.
	Recorder defective.	If recorder pin can be moved easily by hand, recorder is defective.	Consult instrument manual.

Appendix A

CLEANING OF G.C. DETECTORS

During the use of the GC instrument, the detector may become contaminated due to column bleed or sample residues. Whenever possible the detector should be cleaned simply by heating the detector block near the maximum temperature limit for the detector. Wrapping glass or asbestos-insulated heater tape around hard to heat areas may aid in cleaning the detector and prevent future build-up of contamination. Caution: Never heat detectors containing radioactive sources above limits set by the Atomic Energy Commission. Excessive heating may also damage teflon insulators.

If heating fails to remove contamination it will be necessary to remove the detector and clean it using suitable solvents. An ultra-sonic cleaner will be helpful in cleaning many parts of the detector, but generally a few common reagents are sufficient to clean them.

Procedures are given here for cleaning thermal conductivity detectors, flame ionization detectors, and electron capture detectors. These procedures may be applicable to other similar detectors, but care should be taken to prevent the cleaning solvents from damaging detector parts. It is important that clean detector parts not be handled with the fingers. Lint-free linen gloves and tweezers should be used to prevent recontamination of the detector. After cleaning, the detector should be returned to the GC instrument and maintained at operating temperatures overnight before use.

THERMAL CONDUCTIVITY DETECTORS

The following procedure should be used when cleaning the TCD detector:

1. Disconnect all electrical connections except detector block heater.
2. Cap off the outlet port of the detector and fill the detector with Decalin (Analabs. Inc., New Haven, Conn.) through the inlet port.
3. Allow the Decalin to remain in the detector at 100°C for 15 min., then drain.
4. Repeat this procedure three times or until the Decalin comes out clear.
5. Substitute Dimethyl Formamide for the Decalin and repeat steps 2, 3 and 4.
6. Repeat steps 1, 2 and 3 using Methanol at 60°C.
7. Repeat steps 1, 2 and 3 using water at 95°C.
8. Repeat steps 1, 2 and 3 using Acetone at 55°C.
9. Repeat the entire cycle if necessary to completely clean the detector.
10. Pass carrier gas through the detector for about 20 min. after draining it of all solvents and before elevating the detector to operating temperatures.
11. Allow detector to remain at operating temperatures overnight before using.

FLAME IONIZATION DETECTORS

The following procedures are recommended for cleaning flame ionization detectors:

A. Cleaning detector while in place in the GC instrument (for slight contamination).
   1. Remove separation column and insert clean tubing between injection port and detector.
   2. Maintain detector temperature and column oven above 125°C.
   3. Inject three 10μl samples of distilled water.
   4. Inject three 10μl samples of Freon 113 (TM, Union Carbide, Cleveland, Ohio).
   5. Allow detector to remain at 125°C for one hour and check for baseline stability.
   6. If contamination is still present, repeat the procedure or disassemble detector for cleaning as outlined below.

B. Cleaning disassembled detector (for heavy contamination).
   1. Turn off hydrogen and air supplies and disconnect

all electrical connections.

2. Cool detector heater. (Cooling time can be shortened by blowing cool air directly on the detector.)
3. Cap off hydrogen and air inlets.
4. Remove upper flame assembly: barrel, electrodes, and flame jet. (Consult instrument manual for proper disassembly procedure.)
5. If flame jet is made of quartz, clean it in an aqua-regia solution. If the flame jet is made of stainless steel, polish it with #500 emery paper.
6. Wash upper flame assembly (including jet) three times in an ultrasonic cleaner using a 50:50 volume mixture of methanol and benzene (Caution: Vapors are both toxic and flammable).
7. Finally, clean these parts in pure methanol and dry in an oven at 80°C. (Be sure to avoid recontamination during drying.)

   Note: Do not allow halogenated solvents to come into contact with the thermal insulators (teflon, TM) since these will react causing noise to develop upon reassembly.
8. The transfer lines between the column and the flame jet can be cleaned by passing a 50:50 volume mixture of methanol and benzene through them. Continue flushing the transfer lines until the solvent emerges clear, then flush again.

   Note: The methanol/benzene solvent is effective for removing most common contaminants. Use of unusual samples or liquid phases may indicate other solvents might be more effective to remove the contamination.
9. Reassemble the detector, being careful to avoid recontamination, and pass dry carrier gas through it for at least twenty minutes before attempting to heat the detector.
10. Before igniting the flame, allow the detector to remain at operating temperatures (125°C or above) for several hours.

ELECTRON CAPTURE DETECTOR

Upon obtaining the necessary AEC authorization, the ECD can be cleaned as follows:

1. Carefully disassemble the detector, handling the radioactive foil (H^3 or Ni^{63}) with forceps and gloves.
2. Clean metal and teflon detector parts (excluding the

radioactive foil) in a solution of 2 parts H_2SO_4, 1 part HNO_3, and 4 parts H_2O. Use an ultrasonic cleaner if available. Repeat cleaning until solutions are clear.

3. Rinse all parts thoroughly and repeatedly in distilled water, then acetone.
4. Dry in a column oven at 80°C.
5. Specifically authorized licensees may clean the foil as follows:
6. Tritium (H^3) foil can be rinsed in hexane or pentane but *never* in water. The solvent should be discarded with large amounts of water (consult AEC rules and regulations title 10, CRF 20).
7. Nickel (Ni^{63}) foil should only be handled with six inch or longer blunt forceps and should never come into contact with the skin. Either ethyl acetate rinsed with sodium carbonate, or benzene can be used as solvents. The foil should then be immersed in boiling water for five minutes. While AEC regulations should once again be consulted, these solvents can usually be rinsed down the drain with large amounts of water.
8. Reassemble the detector and blow dry carrier gas through it before heating to operating temperatures.
9. Allow the detector to remain at operating temperatures for several hours before use.

Appendix B

CARE AND HANDLING OF MICROSYRINGES

An easily overlooked aspect of the operation of a chromatograph is the use of the injection syringe. This initial operation can have as much effect upon the instrument performance as any other system. Since microsyringes deal in very small quantities (1-10 μl), it is essential that the syringes be kept clean. Cleaning must be accomplished with great care since most microsyringes are quite fragile and their accuracy can easily be destroyed. Be sure to consult the instructions *before* disassembling the syringe rather than afterwards to find out what was done wrong. The technique used for injecting the sample is also important. It is the primary objective of the injection to introduce a sharp "plug" of sample into the chromatograph inlet. Improper injection can cause serious loss of column efficiency. However, by following the few procedures outlined below and with sufficient practice, the analyst will receive repeatedly accurate service from most microsyringes.

CLEANING

Always be sure to clean the syringe before and after use to prevent contamination of samples. If high molecular weight samples are being injected, it is advisable to flush the syringe immediately after use with a suitable solvent to prevent build-up of hard to remove deposits. For general cleaning of an assembled syringe, the following method will usually be successful:

A. Flush the syringe several times with the following solutions in the order indicated:

   1. 5% potassium hydroxide/water
   2. Distilled water
   3. Acetone
   4. Chloroform

B. Remove remaining chloroform by heating and/or vacuum. A simple device for this purpose is made by the Hamilton Syringe Company, Whittier, California 90608.

LIQUID INJECTION TECHNIQUE

A. Be certain syringe is clean.
B. Draw up sample into syringe and flush syringe with sample a few times.
C. Overfill syringe with sample.
D. Invert syringe (needle up) to allow air to rise.
E. Discharge excess sample to desired sample volume.
F. Examine the syringe for bubbles. If they are present, repeat filling procedures. It is sometimes helpful to discharge sample rapidly from the syringe into the sample a few times in order to remove bubbles.
G. Lightly wipe needle with a lint-free tissue, being careful not to draw sample out of needle.
H. In a quick, smooth motion, insert syringe into septum, inject sample, and withdraw syringe.
I. Flush with solvent if necessary.

Appendix C

TROUBLESHOOTING CLUES FROM THE ROTAMETER

The Rotameter attached to the carrier gas line can provide clues to problems occurring in the flow system.

A. Normal Operation

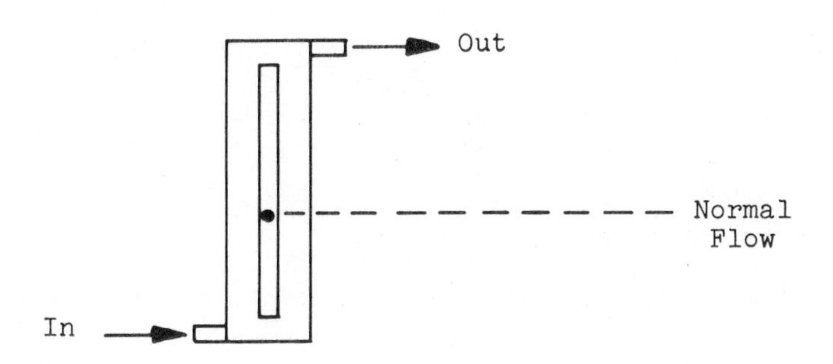

B. Flow System Faults

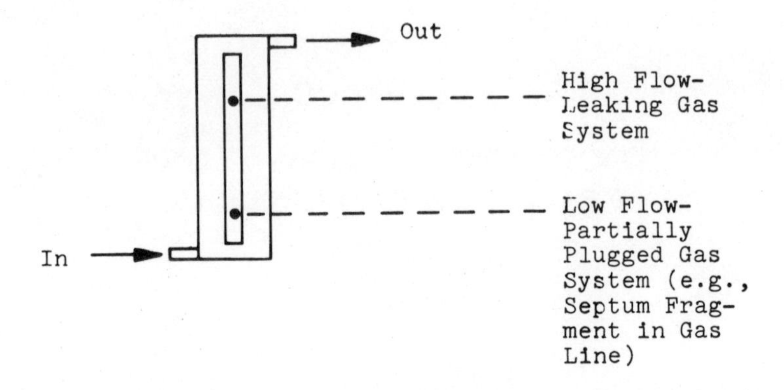

If the Rotameter ball falls below the "Normal" position during a temperature programmed analysis, a faulty flow controller is indicated. The situation may be caused by a low inlet pressure to the flow controller.

Appendix D

TROUBLESHOOTING OF SILICON CONTROLLED RECTIFIER (SCR) HEATING SYSTEMS

The oven and temperature controllers of many modern chromatographs utilize trigger pulse circuits and SCR's to control the power input to the oven heating coils. A typical block diagram of such a circuit is shown in Figure D-1.

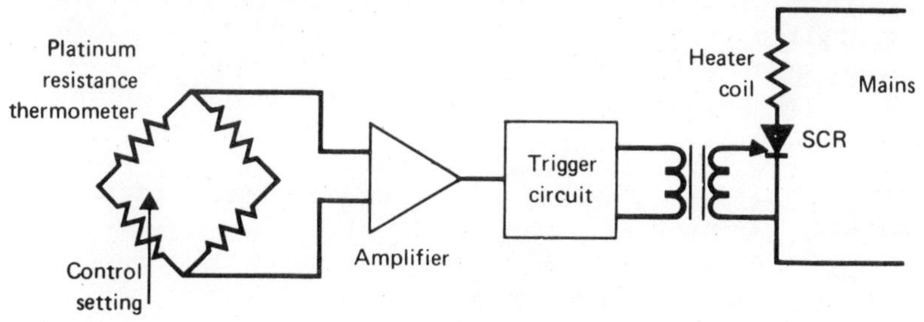

Fig. D-1. Block Diagram of SCR/Trigger Pulse Circuit

The trigger pulse is derived from a resistance bridge, amplifier and pulse shaping circuit. A platinum resistance thermometer inside the oven and the temperature control settings form two arms of the bridge circuit. This signal coupled with a phase indicating signal, often from the secondary coil of a transformer, form the input to the amplifier. The amplifier output is coupled to the pulse shaping circuit which produces the SCR trigger pulse. This pulse varies in phase with respect to the voltage across the

SCR when the oven is in the region of the control temperature and, thus, the power input to the oven is controlled by the out-of-balance signal of the meausring circuit.

The common faults on such a circuit are over or under heating of the oven.

An oscilliscope is invaluable for tracing faults. If the oven fails to heat or fails to heat properly the output from the SCR can be checked. If the output phase patterns, Figures D-2a, 2b and 2c, are not present then the trigger pulse to the gate of the SCR should be checked. If no trigger pulse is present then the measuring circuit and oven controller should be checked for short circuits, open circuits and loose connections. Further attention to the amplifier and trigger circuit is best left to a qualified service man.

If the trigger circuit is present but the oven still fails to heat, the SCR should be checked and replaced if faulty.

The oven overheating situation is usually caused by one of three things. Again the oscilloscope is of great value in tracing the fault.

The first situation is caused by uncontrolled operation of the SCR. This is usually a short across the SCR which must be replaced.

If the SCR is good then the problem is caused by uncontrolled trigger pulses caused by a fault in the trigger generating circuit or, more likely, a shorted platinum resistance thermometer.

The third situation occurs with a phase reversal where the circuit provides more heat instead of less heat. This can lead to an uncontrolled runaway. The usual initial cause is the replacement of a plug or another connecting cable.

Occasionally an intermittent heating condition can arise. This is sometimes caused by incorrect connection of the controller to the oven power supply giving an out-of-phase condition. However, the most frequent cause of this condition is a loose connection.

Oscilloscope set differential across heater leads:

a) Oven temperature below set point

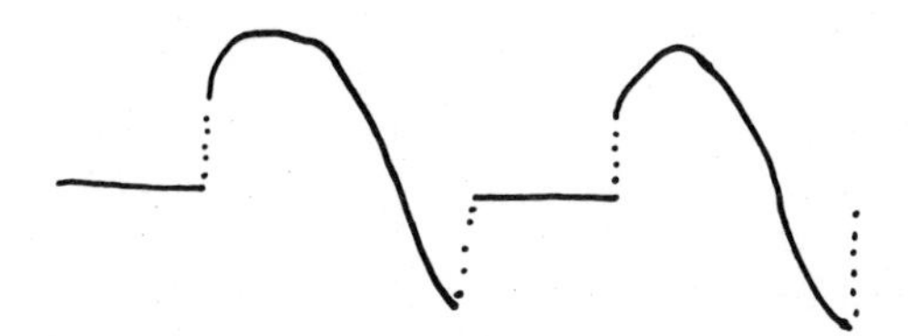

b) Oven temperature close to set point (Normal)

c) Oven temperature above set point

Fig. D-2. Typical oscilloscope SCR patterns.

Appendix E

A RAPID FAULT ISOLATION METHOD

Rapid isolation of faults within the GC system can be a valuable aid in troubleshooting. The method offered here utilizes two simple checking circuits and a dummy column. Essentially, the checking circuits and dummy column are simple, temporary replacements for actual GC systems which might be at fault. The first checking circuit can act as a replacement electrometer, dial in theoretical peaks, act as a replacement for hot wire and thermistor TC detectors, or check out the operation of digital integrators. The second checking circuit acts as a calibration device to check for the proper operation of the strip chart recorder. Faults can be isolated to the column or inlet system by installing the dummy column.

A single electrometer-hotwire-thermistor detector (EHTD) checker can be built from the wiring diagram in Figure E-1. All resistors should be precision and matched and the circuit mounts should be isolated from the case. A series of resistors should also be attached to a terminal strip to be used as replacements for TCD elements. The resistors should match the resistances of the filaments or thermistors and should be mounted in groups of two (for two-element detectors) or four (for four-element detectors) to allow for replacement for all detector elements simultaneously.

The second checking circuit, the recorder checker can be built according to the wiring diagram in Figure E-2.

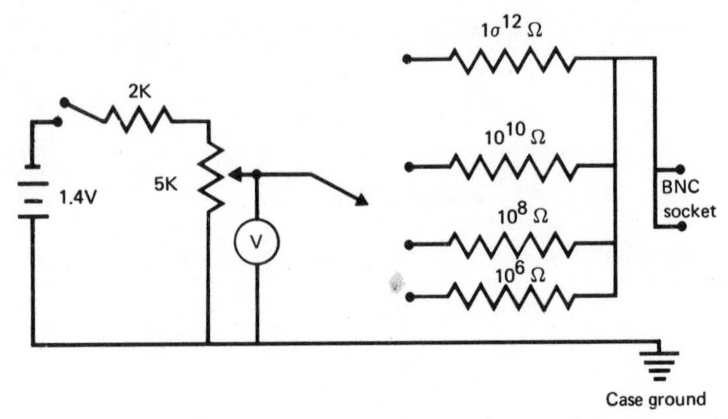

Fig. E-1. Electrometer/hot wire/thermister detector checker wiring diagram.

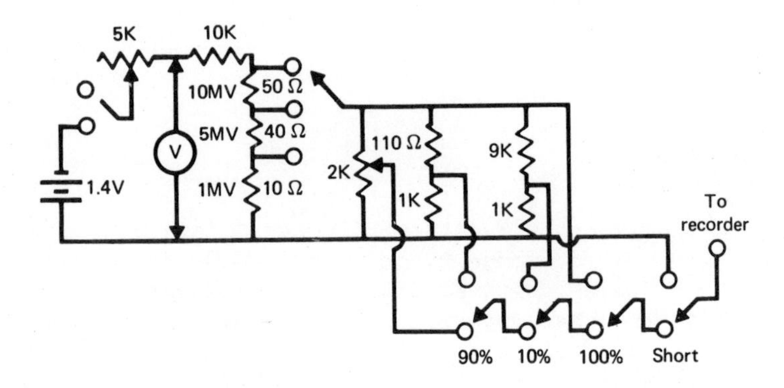

Fig. E-2. Recorder checker and calibration wiring diagram.

By connecting this checking circuit directly to the recorder, it can be easily checked for calibration. The series of switches will cause the recorder pen to travel 90%, 10% and 100% of full scale when they are depressed. This recorder checker is useful in calibrating 1, 5 and 10 mV recorders to within ±2% accuracy. Calibration adjustments should be made according to the instrument manual.

The dummy column consists of a 10-foot length of 1/4" o.d. tubing filled with a molecular sieve support. It is simply substituted for the actual separation column to determine if the fault disappears.

The two checking circuits and the dummy column can be used to isolate faults when used according to the diagram in Figure E-3. This diagram provides a logical method of isolating the fault. It should be noted that this fault isolation method is not intended as a substitute for

Chapter 14, "Gas Chromatographic Comprehensive Troubleshooting." They should rather be complimentary.

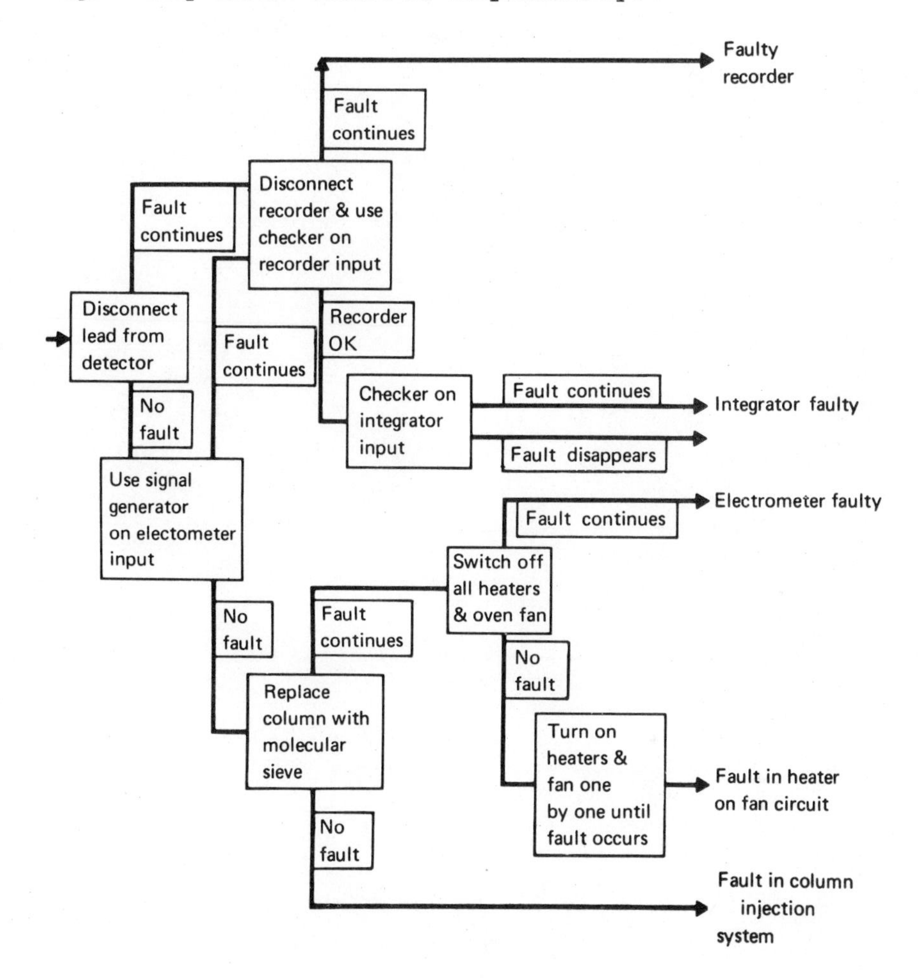

Fig. E-3. Rapid fault isolation diagram.

Appendix F

GAS CHROMATOGRAPHIC COLUMN PREPARATION

Preparing one's own columns, rather than buying commercially prepared columns, can provide a significant cost savings without loss of column capability. Column preparation does not require a great deal of equipment and the materials are inexpensive when compared with the cost of a commercially prepared column. Preparing GC columns does, however, require time and care. If a significant number of columns are needed, the investment in time can be very worthwhile. The methods presented here for preparing both packed columns and open tubular columns have been used successfully to prepare many columns. The methods are not entirely fool-proof though, and a certain degree of failure (relatively inefficient columns) can be expected.

Once the column dimensions have been chosen and the correct liquid phase determined, preparation of the column can begin. All coating procedures are intended to produce uniform packing without voids which will cause losses of efficiency.

CLEANING COLUMN TUBING

As received, tubing, particularly metallic tubing (copper and stainless steel), often contains residues from its manufacture. A cleaning procedure suggested by Gouw, Whittemore, and Jentoft (1) has been quite effective for removing these residues. Cleaning is accomplished by successive flushings with the following reagents:

1. Methylene Chloride (CH_2CL_2)
2. Acetone (C_3H_6O)
3. Water
4. Concentrated Nitric Acid (HNO_3) or concentrated hydrochloric acid (HCl)
5. Water
6. Ammonium Hydroxide (NH_4OH)
7. N-methyl Pyrrolidone
8. Acetone
9. Liquid phase solvent

In each case, flushing should be continued until the reagent emerges clear. Hydrochloric acid should be used with copper tubing in place of nitric acid since HNO_3 will react with the copper. These reagents should be pushed through the column with an inert carrier gas such as nitrogen or helium. Following cleaning, the tubing should be dried with the carrier gas stream for about twenty minutes. The tubing should then be capped under a positive carrier gas pressure to await packing.

COLUMN PACKING PREPARATION

The coating of the liquid phase on the inert support material must be carried out carefully to prevent crushing of the support material and to obtain complete, uniform coatings.

High boiling solvents are preferred when dissolving the liquid phase. A solvent such as toluene would be preferred to such materials as methylene chloride or acetone (2). Using high boiling solvents will reduce the rate of evaporation and provide more uniform coatings.

In the evaporation technique of preparing column packings, the liquid phase/solvent solution is thoroughly mixed with the inert support. Stirring should be accomplished gently with a blunt instrument to avoid fracturing the support material. This slurry is then poured into a shallow evaporating pan and the solvent allowed to evaporate. The evaporation should take place slowly with constant stirring. If the slurry is not stirred during the evaporation process, liquid phase will migrate to the top resulting in high concentration of liquid phase there.

The evaporation technique is useful for 10-30% liquid

phase columns. At percentages less than 10%, the coatings must be more uniform than this method can provide. For these low-liquid-loaded columns, a filtration technique should be used. After the slurry has been formed, it is poured into a glass frit funnel, and the coating solution allowed to drain off. The % liquid phase will be approximately equal to the % of liquid phase in the solution. More accurate determinations can be made by weighing the support before and after coating (3). Once again, it is important to stir the support constantly during drying to prevent migration of the liquid phase to the top.

PACKED COLUMN PREPARATION

After the column tubing is cleaned and the support material coated, the column is ready to be prepared. Columns longer than ten feet should be prepared in increments and joined later with unions. Longer columns are difficult to handle. If the column is six feet or less, it should be coiled prior to packing to prevent crushing of the support during coiling.

Liquid phase and support are packed into the column with the aid of a vacuum pump and sometimes a vibrator. First, a plug of glass wool is inserted into the outlet end of the column. A vacuum pump is then attached to this end of the column to provide a light pull on the column packing. The glass wool plug keeps the packing in the column. Packing is poured into the column through a funnel at the column inlet. For columns which are packed uncoiled, a hand-held vibrator can be used to improve the packing. The vibrator should be started at the outlet end. This can be repeated several times. After the column is completely packed, another glass wool or quartz wool plug should be inserted into the column inlet. The column is now ready for conditioning.

COATING OPEN TUBULAR COLUMNS

Preparation of open tubular columns is somewhat more difficult than preparing packed columns even though fewer steps and less equipment is involved. The columns are prepared by pushing a coating solution through the narrow bore tubing and the difficulty arises in obtaining a thin uniform coating of liquid phase on the column walls. The procedure described here is after *The Dynamic Method* by Ettre (4)

A thoroughly cleaned reservoir capable of containing more than the volume of the column is needed. One end of the reservoir is attached to an inert gas supply (Nitrogen or Helium) and the other end is attached to a 6 foot x 0.01 inch i.d. pre-column restrictor. The gas supply must be equipped with accurate flow controllers and a bleed valve to help maintain constant flow through the column. The column tubing is attached to the pre-column restrictor at one end and another 6 foot x 0.01 inch post-column restrictor at the other end. At the outlet of the post-column restrictor, a soap bubble flow meter or a water reservoir can be used to monitor the flow of the coating solution.

To coat the column, a measured quantity of coating solution in excess of the column volume is filtered and poured into the reservoir. The pressure from the carrier gas supply is gradually increased until the proper flow rate is obtained. Coating solution should be forced through the column slowly at about 2-5 cm^3/sec volume and linear velocity. The appropriate volume and flow rate should be determined for the column diameter used. As the coating solution leaves the post-column restrictor the flow rate will begin to increase, slowly at first then very rapidly. By reducing the inlet pressure and using the bleed valve to release back pressure, a fairly constant flow rate can be maintained. The coating solution should be collected at the outlet to determine the average coating thickness of the liquid phase (based on the total area of the internal column walls).

After the coating solution has completely emerged from the column, a small flow of carrier gas should be passed through the column for one hour. The column is now ready for conditioning.

COLUMN CONDITIONING

All columns must be conditioned before use to remove traces of coating solvents and other impurities which might interfere with an analysis. The column should be placed in a column oven with the inlet connected to the carrier gas supply and the outlet left to discharge into the atmosphere. With gas flowing through the column, the temperature should be increased slowly. The column temperature should be maintained somewhere above the temperature at which it will be operated and allowed to remain at that temperature overnight. Never condition the column for this length of time at temperatures within 20°C of the maximum operating temperature of the liquid-phase. After conditioning, the column is ready for use.

REFERENCES

1. T. H. Gouw, I. M. Whittemore, and R. E. Jentoft, *Anal. Chem., 42*, 12 (1970).

2. Short Course on Column Selection, On Tape, Supelco., Inc., Bellefonte, Penna., 1971.

3. S. F. Spencer in *Instrumentation in Gas Chromatography*, J. Krugers, Ed., Centrey Publishing Company (Eindhoven), 1968).

4. L. S. Ettre, *Open Tubular Columns in Gas Chromatography*, Plenum Press (New York), 1965.

Appendix G

GLOSSARY OF TERMS AND FORMULAS

Absolute Retention Time: (t_R) See, Retention Time.
Absolute Retention Volume: (V_R) See, Retention Volume.
Adjusted Retention Time: (r_R') Absolute retention time (q.v., retention time) minus retention time of a substance which does not interact with the stationary phase (i.e., air, for many columns). Units: usually in millimeters of chart, though it can be given in units of time.
Adjusted Retention Volume: (V_R') Absolute retention volume (q.v., retention volume) minus gas holdup. Units: cm^3
Adsorbent: Stationary phases which interact with sample components only at the surface of the phase.
Air Peak: The peak of any substance which does not interact with the stationary phase (for many columns this will be air).
Attenuator: The part of the electrical system which determines the amount of amplification of detector signals being sent to the recorder.

Baseline: The straight line drawn by the recorder pen when no sample is entering the detector.

Capillary Column: A small inner diameter column (0.25-1.0mm) which contains a stationary phase attached to column walls either directly or on a solid support.

Carrier Gas: An inert gas which pushes the sample through the column. Common carrier gases are helium, nitrogen, argon, and argon/methane.

Carrier Gas Velocity: (μ) Mass flow rate of the carrier gas. Units: ml/min.

Chromatogram: The series of peaks produced by the strip chart recorder representing the sample.

Column: The tubing (metal, glass, or teflon) containing the stationary phase.

Detector: The device which converts the presence of the sample component at the end of the column into an electrical signal. The detector monitors changes in some physical property of the gas.

Elution: The emergence of a component from the column.

Efficiency: (N) The ability of a column to maintain sharp (narrow) peaks. It is the primary criteria for judging column quality.

$$N = 16\ (t_R')/w$$

t_R', adjusted retention time (cm)

w, peak width at base (cm)

Flow Programming: Predetermined increase in carrier gas flow rate to expedite elution of high boiling components without raising column temperature.

Gas Holdup: (Vm) Volume of carrier gas which passes through the column in order to elute a substance which does not interact with the stationary phase.

Gas-Solid Chromatography: Gas chromatography in which the stationary phase is a solid adsorbent.

Ghost Peaks: Unexpected chromatogram peaks usually caused by contamination of the GC system or the sample.

Height Equivalent to a Theoretical Plate: Column efficiency as related to column length.

$$HETP = L/N$$

L, column length (cm)

N, efficiency

Note: The reciprocal of HETP (1/HETP = N/L) is frequently used to describe efficiency in plates per unit length.

Integrator: An instrument used to determine the area under a chromatographic peak. The measurement is either made mechanically or electronically providing a continuous digital read-out.

Ionization Detectors: Detectors which measure current produced by the migration of ionized particles in the carrier gas stream. The particles are ionized usually by either heat or by a radioactive source.

Katharometer: See, thermal conductivity detector.

Leading Peaks: Peaks characterized by a gradual rise of the recorder pen followed by a rapid return to baseline.

Mean Carrier Gas Velocity: (μ) Carrier gas velocity over the length through the column

Mobile Phase: The carrier gas which pushes sample components through the column.

Open Tubular Columns: See, capillary columns.

Pyrolysis: The thermal degradation of a sample before introduction into the column

Relative Retention Time: (r) A measure of the capability of a stationary phase to separate two components. It is the ratio of the adjusted retention times for the two components.

$$r_2 - 1 = t_{R'_2}/t_{R'_1}$$

Note: r is also known as the Separation Factor (α)

Resolution: (R) The degree to which two peaks are separated. It is a function of peak width and distance between the peaks.

$$R = V_{R_2} - V_{R_1}/w_1 + w_2$$

when R = 1.5, separation is complete.

Retention Time: (t_R) The time required for a substance to pass through the column. It is measured from the time of injection of the sample to the time at which the top of the peak appears. Units: cm or units of time.

Retention Volume: (V_R) The product of the retention time (min) and flow rate (ml/min.). Units: ml.

Rotameter: Device for measuring gas flow rates consisting of one or more small balls contained in a length of glass tubing. The flow rate is a function of the height of the ball in the tubing.

Selectivity: The tendency of a detector to respond to certain classes of compounds to a greater degree than other classes of compounds.

Separation Factor: (a) See, Relative retention time.

Soap Bubble Meter: A device for accurately measuring low flow rates. It consists of a volumetric glass tube with a soap reservoir through which the gas is flowing. The flow rate is determined by the time required for a soap bubble to move between calibration marks.

Solid Support: A solid material upon which the liquid phase is coated.

Solvent: As applies to GC, it generally refers to the liquid phase. It may also indicate the substance used to dilute or transport the sample.

Specificity: See, selectivity.

Splitter: A fixed or variable device for reducing the quantity of sample entering the column from the inlet.

Stationary Phase: The liquid or solid material in the column which causes separation to occur.

Tailing: A peak characterized by a rapid rise to its peak, followed by a gradual return to the baseline.

Temperature Programming: Increasing column temperature at a predetermined rate to enhance separation and speed elution of high boiling components.

Theoretical Plates: (N) Unit of column efficiency (q.v.) Height Equivalent to a Theoretical Plate.

Thermal Conductivity Detector: A GC detector which measures changes in the conductivity of the carrier gas caused by the presence of sample components.

Vapor-Phase Chromatography: Synonymous with Gas Chromatography.

Appendix H

PACKING AND INSTALLATION TECHNIQUE FOR G.C. COLUMNS

PROPER PACKING TECHNIQUES

1. Factors to Remember

   1.1 Vacuum. Metal and glass coiled columns are most easily filled with the detector end of the column connected to a vacuum source (10 to 25 lbs/in^2). A trap should be used between the column and vacuum source to catch any packing or glass wool that may be pulled through. A glass wool pad is usually placed between the vacuum hose and the detector end of the column while packing the column.

   1.2 Vibration. A Burgess vibrator assists in packing both metal and glass columns. Dry packing is poured into a dry column through a funnel attached to the injection end of the column. The vibrator should initially be placed at the base of the funnel and turned on low while packing is poured into the funnel. The vacuum will pull the packing into the column while the vibrator maintains a stream of packing flowing into the column. When the column appears full the vibrator is run up and down the column (metal or glass U-tube) or around the column (glass coil). Use the following settings:

Support	Friability	Vibrator Setting
Chromosorb P	Moderate	Low to Medium
Chromosorb W	High	Very Low
Chromosorb G	Very Low	Medium to High
Chromosorb T	Agglomerates	Medium to High
Glass Beads (Corning)	None	Medium to High
Porous Polymers	Very Low	Medium
Adsorbents	Moderate	Low to Medium

1.3 Column Preparations

1.3.1 Metal. Flush the smoothly, pre-cut length of metal with water and then acetone (A.R.). Pull vacuum on the tubing and air dry. Seal tightly with plastic column caps if not packed immediately.

1.3.2 Glass. Work with glass at or slightly above room temperature. Cold glass tends to snap easily, even when tempered. Flush the column with acetone (A.R.), pull gently with vacuum (use a trap) and dry in an oven for 5 to 10 minutes at about 100°C. Cap tightly if not packed directly.

1.4 Column Packings. A reminder to use dry packings may sound elementary, but often adsorbed moisture can cause a problem. Pour the packing into a tared beaker and place in an oven* at the solvent boiling point** for 10 minutes. Cool and pack into the column.

1.5 Glass Wool. Use silanized glass wool to avoid any possible contamination from the glass wool plug. A small amount to hold the packing in the column is all that is necessary.

---

*Teflon is packed cold (0 to 4°C) with the column, funnel and packing all being chilled.

**The solvent used should have a boiling point at least 10° to 20° below the liquid phase.

1.6 Calculated Amount of Packing. A column may be considered a cylinder for calculation purposes.

$V = \pi r^2 h$ e.g., r = 2 mm or 0.2 cm
h = 4 feet or 1.2 m (120 cm)
$V = (3.14)\ (0.2)^2\ (120) = 15.1\ cm^3$
W = DV
Packing density of Chromosorb P = .47g/cc
Chromosorb P x wt. (g.) = 0.47 x 15.2 = 7.1 grams of Chromosorb P (packing) to fill a 1.2 meter x 2 mm (i.d.) x 4 mm (o.d.) column.

2. Packing Techniques. Always weigh the amount of packing used for a column and maintain this record.

2.1 Glass U-Tubes. Connect the arms of the column to a ring stand with clamps. Pour packing through a funnel until each arm is filled. Tap each arm gently and fill again to equal heights. Plug each end with a glass wool.

2.2 Glass Coils. Connect a vacuum source through a trap to the detector end of the column. Place a glass wool plug between the column and vacuum hose. Pour packing into the open end through a funnel. Use vibration as discussed in Section 1.2. Run the vibrator around each coil with the vacuum connected to pack evenly. Plug each end with glass wool.

2.3 Metal. Connect one end of the straight metal tubing to a vacuum source through a glass wool plug. Pour the packing in with vibration near the funnel. Run the vibrator up and down the tubing to produce a uniformly packed column. Since the visual results are not available for metal columns, keep a record of the weight of packing used.

2.4 Teflon Packings. Connect the column or tubing to vacuum (20 to 25 lbs/in^2) through a small glass wool plug. Pour the chilled packing through a chilled nylon screen (10 to 20 mesh sizes greater than packing mesh size) into the chilled funnel. Clamp the vibrator (medium speed) to the neck of the funnel. Fill the column and plug with glass wool.

3. Installation

3.1 Metal. Use S/S fittings for S/S columns, brass for copper and aluminum. Slide the nut, back and front ferrules onto the column. Connect the injection end and snug hand-tight. After conditioning the column, connect the detector end and tighten all fittings with a wrench. Do not over-tighten and crimp the tubing.

3.2 Glass. Install a brass nut and back ferrule and two Viton O-rings on each end. Install the injection side and hand tighten. Condition the column. Install the injector end. Snug the fittings with a wrench. Viton O-rings are the most reliable type. Column temperatures of 250° to 300°C (isothermal) require tightening of fittings daily and O-ring replacement every 6 to 10 runs.

4. Column Storage. Once packed, columns should be capped to prevent moisture pick-up when not in use.

APPROXIMATE PACKING REQUIREMENTS

Packing Type	Packing Density Grams/cc.	Weight Packing (grams/Meter[d]) vs. Column I.D. (mm)				
		1[a]	2[b]	3	4	5[c]
Chromosorb P	0.47	3.0	6.0	9.0	12	15
Chromosorb W	0.24	1.5	3.1	4.5	6.0	7.5
Chromosorb G	0.58	3.7	7.4	8.1	14.8	18.5
Chromosorb T (Teflon 6)	1.0	6.4	12.7	19.2	25.6	63.5
Glass Beads (Corning)	1.4	8.9	17.8	26.7	35.6	43.5

[a] 1/16" metal.

[b] 1/8" metal.

[c] 1/4" metal.

[d] 1 ft. = 0.3 meters.

Appendix I

CONDITIONING OF G.C. COLUMNS

A column prepared for gas chromatography (gas-liquid or gas-solid) requires a treatment known as "conditioning" prior to its proper use for a specified application. The "conditioning process" usually involves a flow and temperature treatment of the column, after preparation. Occasionally the procedure will also involve the addition of certain chemicals, e.g., silylation reagents to esterify free hydroxyl groups or basic hydroxides to alkalize a column for certain amine samples.

Conditioning is needed to remove the last traces of solvent used in coating the liquid phase on the support, moisture entrapped in the support, traces of silylation reagents, impurities in the liquid phases (light ends). After considerable usage or high sample concentration, reconditioning by increasing column temperature to the liquid phase limit for several hours is sometimes needed to remove adsorbed solutes or sample matrix contaminents.

The "conditioning process" results in the following advantages: 1) *Reduction of background noise due to column bleed* (solvent, light ends). This is important for sensitive flame (responds to all carbons) and electron capture (10^{-12} to 10^{-9} g. sensitivity requires a stable low bleed column) analysis. The chromatographer can recognize bleed by a noisy, erratic baseline. It should not be confused, however, with a high recorder gain setting which causes erratic pen movement or static electricity which produces sharp spikes (often peak-like) on the chromatograph trace. True column bleed usually causes a positive baseline drift as well as noise. Chromatographs grounded through the 110 V. outlet plug can still show static charges. 2) *Proper column response and stability*. The required resolution and efficiency depend primarily on the packing or liquid phase selected, the carrier flow rate, the column temperature, and the detector. Once these factors have been determined, the column should be "conditioned" according to the sensitivity required and the detector used. The more sensitive the analysis (i.e., detector), the more thorough the "conditioning" must be. Qualitative analysis, where slight tailing is permissible, requires only "conditioning" about 25°C above the working column temperature for 3 to 12 hours, depending on the liquid phase. Quantitative work requiring 3% or better r.s.d. requires "conditioning" at the upper temperature limit of the liquid phase for periods of 6 to 48 hours depending on the liquid phase.

The following general rules apply to proper "conditioning" of GC columns:

1. The conditioning time and temperature depend on:

   1.1 Upper temperature limit of the liquid phase.
   1.2 % coating of liquid phase, e.g., capillary packed columns; 1%, 10%, 20% vs. "conditioning time."
   1.3 The detector sensitivity required and the type of detector used. The higher the sensitivity, the more thorough the conditioning required.
   1.4 The quality of the liquid phase, i.e., "GC grade" less than "technical" grade, e.g., OV-phases less than XE-60, QF-1 (technical grade).
   1.5 The sample polarity (more polar samples require more extensive conditioning."

2. Detector contamination by nonconditioned columns causes sample tailing, possible sample decomposition, detector malfunction. (NEVER CONNECT COLUMN DETECTOR END DURING "CONDITIONING" PROCESS.)

3. Column length and diameter are negligible considerations for analytical columns less than 3 meters. Longer columns or preparative columns require longer conditioning times.

Several reports in the literature suggest preliminary temperature conditioning at zero flow. While this may be useful in some cases, the author does not recommend this method. Without flow, the liquid phase tends to aggregate on the support and forms bubbles of varying liquid phase concentration. This phenomenon has serious effects on Teflon columns, glass bead columns and all low coatings ($\pm 3\%$). A small flow (10 to 20 cc.) should be maintained at any time that the column is heated.

HOW TO "CONDITION" YOUR G.C. COLUMN

1. Technical Grade Liquid Phases. Since most phases are still prepared in large batches and not specifically for G.C. work, a thorough "conditioning" is needed. Connect the column to an inlet flow of 10 to 20 cc carrier gas with the detector end open. Heat the column to 1/4 of the recommended liquid phase temperature limit for 1 to 2 hours. (Remember these temperature limits are experimentally determined ranges where thermal break-down begins to occur.) Raise the temperature to 25°C above the expected operating temperature for about 6 hours for qualitative work, 24 hours for quantitative work. Silicone phases require more conditioning than Carbowax phases, e.g., do not be disappointed if additional conditioning is required after the first try. Do not try to hurry the process since time loss and frustration are the results.

2. G. C. Grade Liquid Phases. OV phases (Ohio Valley Specialty Company), SE-30 and many polyesters are now available in "purified" form. These materials are prepared in small batches for GC work. Thermal stability is better than the technical grade material. The silicones contain fewer light ends. The silicones generally require 1 hour at 100°C plus temperature programming to the upper temperature limit at 5-10°C/min. and 2 hours at the upper limit. Polar samples, e.g., sugars, alcohols, may require overnight conditioning. Polyesters generally are conditioned at 100°C for 2 hours and at the upper limit for about 8 to 12 hours.

   Again, the polarity of the sample will determine the amount of conditioning necessary to reduce or eliminate tailing.

3. Porous Polymers.

   3.1 Porapak Type. Condition with flow for 1 to 2 hours at the upper temperature limit.

3.2 Chromosorb 100 Series. Condition for 1 hour at the operating temperature for low sensitivity work and overnight at the upper temperature limit for quantitative work.

4. Adsorbents. Acid-washed charcoal, silica gel and alumina columns should be heated 3 to 4 hours at 150°C. Molecular sieve columns should be heated at 250 to 300°C overnight.

In conclusion: Spend the necessary time to condition your column and observe longer column lifetime, better stability and less tailing as a result.

Appendix J

ROUTINE MAINTENANCE OF YOUR G.C.

Most laboratories using gas chromatography maintain adequate supplies for normal operation, but have no electronics technician to provide rapid gas chromatography repair service. The operator (chemist, clinician, technician and so on) must usually double as the plumber and electrician for his instruments. The only troubleshooting aids usually available are a manual (who knows how to decipher the circuit diagrams?) and common sense. The hope in the following article is to describe certain hints for daily operation that may reduce downtime and expensive service calls.

The gas chromatographic system (instrument) can be classified into four maintenance areas:

1. Gas-inlet system
2. Column Oven
3. Detector
4. Readout device

1. Gas-Inlet System.

1.1 Carrier Gas. Helium and nitrogen are commonly used for T.C., F.I.D. and E.C. detectors. Purity of these gases should be 99.9%+. Preventive maintenance equals selection of a dependable supplier of pure carrier gas. Molecular sieve and/or silica gel drying tubes are usually unnecessary except for high sensitivity E.C. work or where oxygen and water are being determined in the sample.

Two stage regulators are recommended for the normal GC pressure range of 0-100 (psig). The gas regulators connected to the GC carrier gas inlet by copper tubing (1/8") and brass fittings (nut, back and front ferrule). 1/2" Teflon^R sealing tape should be used to connect (a) the regulator to the tank and (b) the gas line to the regulator and GC carrier gas inlet. Check each fitting connection with a soap solution (SNOOP^R)[1], (Leak-Tek^R)[2] after installation of every tank of gas. Periodically check (weekly) tank and GC inlets for leaks.

1.2 Detector Gases. Contaminated helium or nitrogen (T.C.) wet hydrogen or air (F.I.D.) and impure helium (Beckman Ionization E.C.) or argon/methane (H^3, Ni^{63} E.C.'s) reduce or destroy detector sensitivity. Molecular sieve and/or silica gel drying tubes should be connected between the tank and GC inlet for all detector gases. Teflon tape should be used for all detector gas connections as described under Section 1.1. Check hydrogen connections daily with soap solution. Insist on 99.995+ purity for all detector gases. When the inlet pressure falls below 50 psig, watch chromatograms for ghost peaks that may result from gas contaminants.

SAFETY NOTE: FID Users, remember to turn off H_2 flow when changing columns.

1.3 Inlet Septa. Many problems with leaks, ghost peaks, and reduced sensitivity begin at the septum. Initially one must use a "good" septum. This product is one which:

(a) does not emit contaminating compounds into the GC column at the injection port temperature used and with the solvent/solute mixture being analyzed.

(b) reseals itself to a leak-tight system upon repeated syringe puncture.

Teflon^R-backed septa (Microsep^R) give long lifetimes (up to 200 syringe injections and yield essentially insignificant contamination. Butyl rubber septa give significant background except for operations below 100°C.

Usually the septum should be replaced every 50 to 200 injections. The operator should know how good his septa are and not fail to change them on schedule. Repeated temperature programming can reduce septum lifetime by a factor 5 to 10% if there is some heating and cooling at the septum inlet with each programmed run. Most chromatographs do not have complete insulation between column oven and inlet, so this condition will occur with programming with most GC units.

Septum lifetime can be increased by reducing the inlet temperature to that of the column when on-column injection is used. The thought that the injection port temperature must be 25 to 50 degrees above the column is untrue for on-column injection.

Summarizing:

<u>Gas-Inlet Maintenance Checks</u>

1. Use pure carrier and detector gases.
2. Use Teflon[R] tape for all fittings.
3. Check all gas fittings (weekly) or (daily) for leaks.
4. Change septums on a schedule empirically determined by the number of injections and the injection port temperature.

<u>G.C. Malfunctions That May Be Related to Gas-Inlet Leaks</u>

1. Loss of signal (major leak) or reduced signal (slow leak).
2. Retention time increases.
3. Peaks become diffuse or spread-out.

Note: A leak in the carrier gas system may result in column packing burn-out and thus loss of the column. Many column packings decompose during "dry heating" (no flow), especially when the column temperature is near the upper limit for the liquid phase.

2. <u>Column Oven</u>.

Most problems with the column oven occur because the column packing is abused, column fittings are leaking, columns are not changed when necessary and fan motors are not cared for.

Column packings must be conditioned properly prior to use to remove excess solvent used during preparation of the packing and "light ends" (low molecular weight liquid phase materials or contaminants). Metal columns should be fitted with proper nuts (S/S for S/S columns, brass for aluminum, copper, teflon columns) to prevent galling of fittings on the tubing and provide leak-tight seals. As a general rule, tighten all metal column fittings once a week or more often if programming is used. Glass columns require brass nuts, brass back ferrules and two silicon or viton O-rings (for below 200°C) or Teflon front ferrules (200-300°C) for adequate seals. Teflon O-rings appear to be too rigid for easy hook-up of columns. The fittings should be tightened by hand during conditioning and then snugged with a wrench prior to actual runs. Repeated programming puts stress on the glass column seal and necessitates more frequent changing of the O-rings or Teflon front ferrules. The best rule to follow is to observe the retention time of a given standard for any increase. When this occurs at constant column temperature (and no other leaks are apparent) the fittings need replacing. Isothermal operation at or below 275°C might reduce O-ring or front ferrule changes to only every 2 to 3 weeks. A small tip: when changing column temperature--low to higher (go at 25°C increments) higher to lower (if possible, change temperature controller to lower temperature and allow to fall slowly without opening the oven door). These methods reduce the need for glass column seal replacement. Fittings on glass columns should as a rule be tightened daily, if possible. Leak check all fittings when installing metal or glass columns and again after conditioning. Glass columns have the advantage of showing when the packing material has started to deteriorate or become contaminated by sample. When the packing appears to have darkened or charred, the column should be removed. In many cases the darkened portion can be replaced by new packing and the column reconditioned for 2-3 hours before returning to normal runs.

Most mechanical problems with column ovens occur because the blower fan begins to loosen and vibrate. The flywheel then vibrates causing increasing torque and strain on the motor shaft. This condition eventually leads to motor burn-out. There is no reason for a motor noise other than a low hum. Investigate immediately any vibration or louder noise that comes from the column oven. An occational tightening of motor fanbolts or stays can eliminate a costly repair bill.

Summarizing:

Column Oven Maintenance Checks

1. Use proper fittings for metal and glass columns.
2. Check fittings with soap solution upon installation and post-conditioning, (use high temperature Leak-Tek^{R}).
3. Tighten fittings weekly (metal) or daily (glass) as per column tubing used.

4. Change columns or replace deteriorated packing upon poor appearance (glass) or poor performance (glass, metal).
5. Discontinue operation and eliminate motor fan vibrations.
6. Check column temperature calibration device occasionally by adjusting the pyrometer reading and temperature control dial to read the same at three different temperatures (75, 150, 275°C).

G.C. Malfunctions That May be Related to:

Column Leaks

1. No sample peaks
2. Increasing retention times
3. Peak tailing or spreading
4. Decreased sample sensitivity

Column Deterioration

1. Increased column bleed (noisy or increasing baseline signal)
2. Reduced sample component resolution
3. Decreasing sample sensitivity
4. Decreasing retention time
5. "Ghost" peaks due to bleed

Glass and metal inlet liners should be removed periodically for cleaning. Use acetone, ethanol and 5 to 10 minutes baking out to remove adsorbed materials. Columns with one piece construction through-to-the-septum should be removed and swabbed with acetone and ethanol occasionally. The need for this type of maintenance will depend on the sample type, its viscosity, thermal decomposition, nonvolatile components, and so on. Use A.R. grade acetone and ethanol.

3. The Detector

Most chemists know the least about the construction and electronic functioning of their detector(s) than other parts of their GC. Therefore, maintenance of a close check on detector performance, realization when a malfunction is apparent and calling the manufacturer's repair service are the best policies.

3.1 How to Treat Various Detectors

3.1.1 Thermal Conductivity:

a. Always purge with carrier gas before switching on the voltage.
b. Turn off the T.C. power when not in use.
c. Reduce the current (ma) to the lowest value that will give the desired sensitivity.
d. Occasionally turn the detector temperature up to remove adsorbed materials.

3.1.2 Flame-Ionization Detectors

a. Determine the carrier/hydrogen ratio that gives the best general sensitivity for your compounds and stick to it (usually 3/2). This is conveniently done by running a standard and varying the H_2 flows vs. a fixed carrier flow.
b. Operate at 10-15°C above column temperature to prevent compound condensation.
c. Occasionally (over the weekend is fine) adjust the detector temperature to 25°C above that temperature used normally to "bake" off the adsorbed materials. This is essentially important when chromatographing biological, viscous or reactive samples.
d. Clean flame jets with a swab soaked in acetone/ethanol (1.1) weekly. Clean flame jet covers also. The latter are frequent sources of detector noise since carbon residue from solvents (especially chlorinated hydrocarbons) and samples rapidly blacken the underside of the cover and fall back into the flame.
e. Use "pure" H_2 and dry air, ONLY.
f. Weekly (Friday afternoon is good) inject 10 microliters of Freon 113 and adjust the detector temperature to 25°C above normal for several hours.

3.1.3 Electron Capture

a. Use "pure" argon/methane (Radioactive) or ultrapure helium (Beckman ionization).

b. Determine the correct carrier/detector flow ratio for each compound (E.C. detector temperature sensitivity varies according to structure or "capability"; higher detector temperature does not necessarily mean higher sensitivity).
c. Carefully follow cleaning instructions for contaminated detectors. A reduction in sensitivity, an increase in background noise and an increasing solvent tail are indications of this condition. Ultrasonic cleaning is best. Call a service man for this job.
d. Use only pesticide grade, neutral solvents--hexane, heptane, benzene--unless others are absolutely necessary.
e. Avoid injections greater than 25 nanograms per run. These precautions permit long periods of negligible contamination.
f. Avoid oxygen in the carrier gas (reduces sensitivity) or argon/methane.

3.1.4 Other special detectors such as the gas-density, coulometric, thermionic, phosphorous and so on require similar treatment to the electron capture. Follow manufacturer's guidelines.

3.1.5 General Detector Rules:

a. Never connect the column to the detector inlet until AFTER conditioning.
b. Never handle detector cables with bare hands. Contamination of detector plug-ins is extremely easy. The only cure is new cables.
c. Never apply power to any detector unless a gas flow is purging it.
d. Water in the sample.

T.C. - The correct detector
F.I.D. - 0.1%
E.C. - Never

G.C. Malfunctions That May be Related to Detector Trouble

1. Noisy baseline or reduced signal	1. Loose or short-circuited cable, "dirty" detector.
2. No signal	2. Burned out element (T.C.) plugged jet, improper carrier/H_2 Ratio (F.I.D.) overloaded E.C.
3. Noise "spikes" on chromatogram	3. "Dirty" detector, electronic noise.

4. The Readout System

Our main concern is the recorder. While integrators and computer hook-ups are now routine, we shall leave these to the electronic experts.

4.1 Routine Recorder Checks:

1. Adjust zero and 100% full scale daily.
2. Clean the slidewire weekly. Use the special silicone oils or sprays sparingly. An oily slidewire is worse than a dry one.
3. Adjust gain several times each week. High gain gives noise spikes. Low gain causes pen drag and wear.
4. Flush pen lines at least monthly. This operation will pay for itself in the long run.

4.2 Recorder Operation

1. Leave the power ON, always. Nothing ruins electronic devices faster than frequent power starts and stops.
2. Avoid changing chart speeds in mid-stream. Return to neutral, turn off chart motor and then adjust to a new chart speed.
3. Avoid changing attenuation when a peak (compound) is eluting. This operation overloads the pen circuit since the recorder is now unbalanced. Attenuators and integrators have electronic compensation, but manual attenuation does not.
4. When recorder servo motors blow out, call for service. These devices are not easily repaired in the laboratory.

A Routine Program for Your G.C.

1. Leave the power on at all times, even if you use the G.C. once a week.
2. Adjust the carrier flow to 1/4 the operating rate when not in use.
3. Daily: 3.1 Adjust all flows and temperatures.
   3.2 Inject a known standard sample containing an internal standard. If retention times decrease, sensitivity decreases, or no peaks occur, begin by changing the septum and work forward until you find the trouble spot.

Products for Low Maintenance G.C.

1. Gas-Inlet System

   1.1 Matheson gas regulators
   1.2 Teflon[R] Sealing Tape
   1.3 Brass, S/S and Teflon fittings, O-rings, ferrules.
   1.4 Snoop bubbler meters with magnetic clip.
   1.5 Snoop[R]
   1.6 Leak-Tek[R]
   1.7 Molecular sieve traps
   1.8 Silica gel traps
   1.9 Microsep[R] Teflon-backed septums--all types.

2. Column Oven

   2.1 Pre-conditioned packings and packed glass and metal columns
   2.2 Teflon front ferrules
   2.3 Column wrench set.

3. The Detector

   3.1 Freon 113 cleaning solution
   3.2 Cat whisker set for F.I.D. jet cleaning
   3.3 Oxysorb for elimination of O_2 in gases.

4. Recorder

   4.1 Slidewire spray cleaner (non-oily)
   4.2 Chart paper for all recorder models

Appendix K

GAS CHROMATOGRAPHY OPERATING HINTS

The method used in gas chromatography will vary with the type of analysis and the individual techniques of the operator. The following discussion will present some preliminary operating hints that may help in obtaining faster and more accurate results with a gas chromatograph.

Columns and Packings. All new columns should be preconditioned before being used. This preconditioning requires "baking" the column at about 50°C above the desired operating temperature of the column for at least 12 hours. This helps remove contaminants in the packing material and minimizes contamination of the detector.

Carrier-Gas Flow Rates. The flow rate of carrier gas through the instrument will affect sample retention time and the amplitude of the recorded peaks. Flow rates between 15 and 30 ml/min should offer the best compromise between sensitivity and resolution (1/8-inch OD columns). Flow rates of 40 to 100 ml/min are recommended for 1/4-inch OD columns.

Detector Temperatures. The detector temperature should be held approximately 25 to 50°C higher than the maximum operating oven temperature. This will prevent condensation of sample components in the detector. Because the detector requires a finite time to equilibrate, its temperature should be changed only when necessary.

Injector Temperatures. Injector temperature should be set approximately 25 to 50°C above the highest column-oven temperature. If thermal decomposition results when the sample is injected into the heated injector, on-column injection should be used. To inject a sample directly onto the column, a needle of at least 2 inches in length must be used.

Isothermal Operation. Generally, the column-oven temperature is set about 40°C below the boiling point of the highest boiler in the sample. Depending on the sample, you may improve results by increasing or decreasing the oven temperature. As a general rule, retention time decreases about 5% for every one degree C increase in temperature.

Programmed Operation. Programmed temperature techniques permit the column oven to operate through a temperature range to elute the low boilers in a well spaced pattern and the high boilers in a reasonable period of time.

Sample Injection. Good sample injection techniques are learned with practice. The syringe is held perpendicular to the injector septum to ensure reproducibility. Sample injection must be in a smooth but rapid motion to permit complete vaporization of the sample. This technique will prevent peak tailing due to poor injection. Also, flush the syringe several times with the sample before filling. Solid samples can be dissolved in a solvent having a lower or higher boiling point than the sample's components. This will allow the solvent peak to elute before or following the sample's components. For further discussion, refer to SYRINGE HANDLING TECHNIQUES section.

Preliminary Analysis. The results of a preliminary analysis will aid the operator in determining optimum operating conditions such as sample sizes, temperature parameters, flow rates, and attenuation. If the initial (or isothermal) temperature is too low, the first peaks will be well spaced, but the remaining peaks will be too widely spaced. If the initial temperature is too high, the first peaks will be too closely spaced or overlapped. Therefore, programmed temperature techniques may offer the best solution for obtaining satisfactorily spaced peaks.

Carrier Flow Balancing for Dual-Column Differential Operation. The primary advantage of using dual columns is to minimize the effects of substrate bleeding from the columns during temperature programming. Substrate bleeding occurs when the substrate vapor concentration in the carrier gas becomes high enough to cause signal drift from the detector and the chromatographic baseline. The rate of substrate bleeding increases as the column temperature increases. The following procedures will minimize baseline drift resulting from substrate bleeding.

Column Flow Balancing for Thermal-Conductivity Detector. Proceed as follows:

1) Adjust carrier-gas flow rate through reference and analytical columns to desired value. Measure flow rate at collector-exit tubes.
2) Heat column oven to 15°C above the desired oven temperature limit. Make sure flow rate in analytical column has not changed.
3) If baseline drift occurs, compensate by adjusting reference-column flow rate to return recorder pen to original baseline.

   NOTE: The effects of flow rate changes occur slowly at higher temperatures; therefore, allow enough time for flow system to equilibrate--up to 5 minutes might be necessary.

4) Lower column oven temperature to lower limit.
5) If baseline drift occurs, adjust FINE ZERO control to return recorder pen to original baseline. Again, allow time for flow system to equilibrate.
6) Repeat balancing procedure by adjusting reference column flow rate at high and fine zero at low temperatures until baseline drift is minimized.

Column Flow Balancing for Dual Flame-Ionization Detectors.* Proceed as follows:

1) Balance electrometer as outlined in operating manual.
2) Adjust flow rate of both columns to approximately 25 ml/min for 1/8-inch columns.
3) Ignite flames and turn ionization voltage on.
4) Adjust column oven to desired initial temperature.
5) Heat column oven to 15°C above final operating temperature.
6) If baseline drift occurs, compensate by adjusting the reference column (Column B) carrier-gas flow rate to return recorder pen to original baseline. Column A is used as the analytical side.
7) Cool column oven to initial program starting temperature. Allow flow system to equilibrate.
8) If the baseline has shifted, compensate by adjusting BUCKING control (reference channel) to return recorder pen to original baseline.

*Columns must be preconditioned for 12 hours. Matched columns must be preconditioned at the same temperature and flow rate.

9) Repeat above procedure (5) through (8) to make sure column bleed effects have been minimized.

A temperature satisfactory for operation of a column with a thermal-conductivity detector may not be satisfactory for operation with the flame-ionization detector. Generally, a much lower temperature must be used. The silicone columns normally have very negligible background signals at 300°C.

Background Signal for Different Columns at Various Temperatures and Times X1 Attenuation

Temp. (°C)	A	B	C	D
100	200	116	64	60
150	367	128	75	70
200	592	268	250	95
250	944	400	3000	188

A: 16% Carbowax 20M
B: 5% KEL-F
C: 10% Poly-M-phenyl ether (5 rings)
D: 5% SE-30 Silicone

The background signal of most high polymeric substrates show an exponential increase with temperature. The substrate poly-m-phenyl ether (containing only 5 aromatic rings), however, shows a radical increase in background signal at temperature above 200°C. This type column is not recommended for use with the flame detector at temperatures above 200°C.

Column Cleaning. Over a period of time the column in your instrument can accumulate materials which cause an increasing background signal. This signal can be reduced by a "steam cleaning" process. A few injections of water at moderately high temperatures desorbs the material from the column and reduced this annyoing background signal.

Scrubber Column. Occasionally it is necessary to use a column packing which has a high bleed rate. For example: Dimethyl sulfolane column for separation of lower hydrocarbons in air. A short column of 4 to 6 inches in length packed with a high molecular weight material such as Carbowax 20M may be connected to the end of the main column. This scrubber column will irreversibly absorb the column bleeding yet permit the passage of the sample components with little change in retention times. It will materially reduce the background signal yet permit a good separation.

Effects of Steam Cleaning

No. of water Injections (10 lambda each)	Background Signal at X1 250°C
0	188
1	72
2	40
3	40
4	16

INDEX

P

R

S

T

U

V

W

Z